Cambridge Imperial and Post-Colonial Studies Series

General Editors: Megan Vaughan, King's College, Cambridge and Richard Drayton, King's College, London

This informative series covers the broad span of modern imperial history while also exploring the recent developments in former colonial states where residues of empire can still be found. The books provide in-depth examinations of empires as competing and complementary power structures encouraging the reader to reconsider their understanding of international and world history during recent centuries.

*Titles include:*

Sunil S. Amrith
DECOLONIZING INTERNATIONAL HEALTH
India and Southeast Asia, 1930–65

Tony Ballantyne
ORIENTALISM AND RACE
Aryanism in the British Empire

Peter F. Bang and C. A. Bayly (*editors*)
TRIBUTARY EMPIRES IN GLOBAL HISTORY

James Beattie
EMPIRE AND ENVIRONMENTAL ANXIETY, 1800–1920
Health, Aesthetics and Conservation in South Asia and Australasia

Robert J. Blyth
THE EMPIRE OF THE RAJ
Eastern Africa and the Middle East, 1858–1947

Roy Bridges (*editor*)
IMPERIALISM, DECOLONIZATION AND AFRICA
Studies Presented to John Hargreaves

L. J. Butler
COPPER EMPIRE
Mining and the Colonial State in Northern Rhodesia, c.1930–64

Hilary M. Carey (*editor*)
EMPIRES OF RELIGION

Nandini Chatterjee
THE MAKING OF INDIAN SECULARISM
Empire, Law and Christianity, 1830–1960

T. J. Cribb (*editor*)
IMAGINED COMMONWEALTH
Cambridge Essays on Commonwealth and International Literature in English

Michael S. Dodson
ORIENTALISM, EMPIRE AND NATIONAL CULTURE
India, 1770–1880

Jost Dülffer and Marc Frey (*editors*)
ELITES AND DECOLONIZATION IN THE TWENTIETH CENTURY

Ulrike Hillemann
ASIAN EMPIRE AND BRITISH KNOWLEDGE
China and the Networks of British Imperial Expansion

B. D. Hopkins
THE MAKING OF MODERN AFGHANISTAN

Ronald Hyam
BRITAIN'S IMPERIAL CENTURY, 1815–1914: A STUDY OF EMPIRE AND EXPANSION
Third Edition

Iftekhar Iqbal
THE BENGAL DELTA
Ecology, State and Social Change, 1843–1943

---

**Cambridge Imperial and Post-Colonial Studies Series Series Standing Order
ISBN 978–0–333–91908–8 (Hardback) 978–0–333–91909–5 (Paperback)**
(outside North America only)

You can receive future titles in this series as they are published by placing a standing order. Please contact your bookseller or, in case of difficulty, write to us at the
address below with your name and address, the title of the series and the ISBN
quoted above.

Customer Services Department, Macmillan Distribution Ltd, Houndmills,
Basingstoke, Hampshire RG21 6XS, England

# Infectious Disease in India, 1892–1940

## Policy-Making and the Perception of Risk

Sandhya L. Polu

First published 2012 by
PALGRAVE MACMILLAN

Palgrave Macmillan in the UK is an imprint of Macmillan Publishers Limited, registered in England, company number 785998, of Houndmills, Basingstoke, Hampshire RG21 6XS.

Palgrave Macmillan in the US is a division of St Martin's Press LLC, 175 Fifth Avenue, New York, NY 10010.

Palgrave Macmillan is the global academic imprint of the above companies and has companies and representatives throughout the world.

Palgrave® and Macmillan® are registered trademarks in the United States, the United Kingdom, Europe and other countries.

ISBN 978–0–230–35460–9

This book is printed on paper suitable for recycling and made from fully managed and sustained forest sources. Logging, pulping and manufacturing processes are expected to conform to the environmental regulations of the country of origin.

A catalogue record for this book is available from the British Library.

A catalog record for this book is available from the Library of Congress.

*To my parents*

# Contents

# Acknowledgments

Many scholars and donors have helped make my research and this book possible. Throughout its several stages, I had the unwavering support of my doctoral advisors. Sugata Bose taught me the importance of transnational and international linkages, a lesson that proved critical to the formulation of this work. Charles Rosenberg provided invaluable guidance and an unparalleled depth of expertise on the history of medicine and public health. David Armitage was instrumental in helping me shape my understanding of empires, international relations, and sovereignty and how those concepts have changed over time.

In the very early stages of thinking about health policy in India, Robert Travers showed great enthusiasm for my ideas and was always generous in imparting his knowledge of the British Empire and India. Durba Ghosh helped me navigate the collections of the India Office Records. The part of my research conducted in Geneva benefited from Sunil Amrith's suggestions regarding archives and resources. Maneesha Lal provided insightful feedback in the early stages of my thinking about how international sanitary conventions affected public health policy in India. Mary Wilson taught me to appreciate the numerous factors affecting the epidemiology of infectious diseases. Dean Jamison helped me to understand 'cost-efficacy' in terms of public health, meanings of which were contested in colonial India and continue to be debated today. Manojit Das Gupta at the Indian Tea Association was kind enough to provide me with access to the association's collection of annual reports. I would also like to thank William Rodriguez for directing me to sources on United States HIV/AIDS policies. There have also been many colleagues and friends who have helped along the way. Allyson Field, Penny Sinanoglou, Sana Aiyar, Antara Dutta, Chanchal Dadlani, Hannah Weiss Muller, Aliza D'Amico-Wong, Faisal Chaudhry, Kris Manjapra, and Nico Slate were great listeners and insightful critics as well as invaluable resources for deciphering archives.

This project required many trips abroad to complete my research, and it would not have been possible without the generous support of grants from the South Asia Initiative at Harvard University, the Department of History at Harvard University, the Jens Aubrey Westengard fund, and the United States Department of Education.

I wish to thank the librarians, archivists, and staff at Harvard University Libraries, the British Library, the Wellcome Institute for the History of Medicine, the Public Record Office at Kew, the World Health Organization, the League of Nations Archives at the United Nations, the West Bengal State Archives, the National Library (Kolkata), the National Archives of India, and the South Asia Initiative at Harvard University. I am also indebted to the many professors and staff at the Department of History at Harvard University.

Finally, the journey that led to this book would not have been possible without my family's generosity and support.

# List of Abbreviations

GOB    Government of Bengal

GOBo  Government of Bombay

GOI     Government of India

GOM   Government of Madras

IMG    Indian Medical Gazette

IMS    Indian Medical Service

IOR    India Office Records

IRFA   Indian Research Fund Association

LGB    Local Government Board

LON    League of Nations

NAI    National Archives of India

OIHP  Office International d'Hygiène Publique

WHO   World Health Organization

# Introduction

> The discourse of risk begins where our trust in our security ends and ceases to be relevant when the potential catastrophe occurs. The concept of risk thus characterizes a peculiar, intermediate state between security and destruction, where the perception of threatening risks determines thought and action.[1]
>
> Ulrich Beck

For decades, if not centuries, perceived risk has played a vital role in shaping infectious disease policies. Throughout history, experiences and memories of epidemics, with their often inexplicable and sudden onset and disastrous consequences, have created perceptions of a world beyond human control, and, thus, the perception of a world at risk.[2] The developed world's approach to the spread of HIV/AIDS is one of the more recent and prominent examples of how global and national perceptions of risk influence the timing and content of policy. During the 1990s, the rising incidence and spread of HIV/AIDS garnered world attention, culminating in Secretary-General Kofi Annan's call in 2001 to create a Global Fund dedicated to fighting HIV/AIDS and other diseases.[3] In December of that year, Jeffrey Sachs, the renowned economist, presented his commissioned report, 'Macroeconomics and Health: Investing in Health for Economic Development' to the World Health Organization, in which he stated that the HIV/AIDS pandemic was a 'distinct and unparalleled catastrophe in its human dimension and its implications for economic development'.[4] Unlike many other diseases prevalent in developing countries, which disproportionately kill infants and the elderly, HIV/AIDS kills young adults, striking at the heart of labor forces and hurting national productivity.[5] Sachs argued that throughout the 1980s and most of the 1990s, the world had avoided addressing the HIV/AIDS pandemic, preferring to focus on global trade and finance issues. However, countries and international authorities failed to grasp that trade becomes much less significant of an issue if HIV/AIDS is killing 20 percent of your labor force. Either a global community needed to work to halt the spread of these disease pandemics or the economic crisis, especially in Africa, would continue to worsen.[6] In 2002, the United States National Intelligence Council also issued a report identifying HIV/AIDS as an important threat to American national security. The United States Intelligence Community argued that HIV/AIDS could exacerbate economic, social, and political instabilities in developing and especially in former communist countries and could destabilize governments and societies in countries of strategic importance to the US, particularly Nigeria, Ethiopia, India, China, and Russia.[7] In 2001, the total US contribution to the global HIV/AIDS crisis was $840 million; it reached $2.3 billion in fiscal year 2004 after President George W. Bush launched the President's Emergency Plan for AIDS Relief (PEPFAR), and most recently amounted to $6.61 billion in fiscal year 2011.[8]

HIV/AIDS was by no means the first disease to receive significant funding from a state because of the political, economic, and, of course, epidemiological risk it posed to national interests. Governments' perceptions of the dangers of disease were just as important over a hundred years ago, when epidemics and pandemics were still in historical memory. Long before the terms 'global health', 'international health security', and 'public health preparedness' came into existence, European and colonial governments struggled to contain and prevent the spread of epidemic diseases from India to the Western world.[9] They viewed epidemics as threats to their territorial integrity, even if international discourse was not explicitly framed in terms of 'national security'. International and colonial decisions about epidemic disease policy were essentially security decisions, framed in terms of probable events, costs, and a spectrum of policy choices that could determine a range of outcomes lying between the optimal and the acceptable.[10] Although the terminology had yet to develop, governments well understood these concepts and their interconnectedness.[11]

International health policy in the late nineteenth century was protectionist in nature and dictated primarily by European powers. The significance of India to Europe – commercially, epidemiologically, strategically – meant that India occupied a central position in debates on the control of epidemic diseases, becoming a focus of international concern and regulation. European health authorities viewed India as a permanent reservoir of cholera and plague. Their anxieties grew during the later colonial period as increased global trade and faster transportation heightened the risk that these diseases would spread from India to Europe.

Risk and fear, however, were not limited to European states. The Government of India's recognition of the risk that disease posed to India's health, labor, economy, and domestic and international political relations played a critical role in policy-making at the all-India level. International perceptions of risk also affected policies in India. For example, when, in 1900, many Muslims evaded restrictions on travel to the Hedjaz by giving false addresses to authorities or by claiming to be merchants, Abdur Rahman, Secretary to the Muhammadan Literary Society of Calcutta, asked the government to permit residents of Calcutta and other plague-free districts to go on pilgrimage to the Hedjaz. The Government of India replied that evasions were not a valid reason to modify restrictions; in relaxing any rules, more importance had to be given to how foreign governments would interpret the action than to the actual risk.[12]

The Government of India recognized that infectious diseases posed certain economic, political, and epidemiological risks. The aim of this work is to understand how both the Government of India's and global perceptions of the risks posed by cholera, plague, malaria, and yellow fever influenced India's infectious disease policies, the formation of which was inextricably tied to international debates on the prevention and control of epidemic disease. This book analyzes how a variety of factors and assumptions, including international public health diplomacy, epidemiology, trade protection, imperial governance, new medical technologies, and cultural norms, operated within a larger conception of risk to shape the Government of India's infectious disease policies. It, thus, focuses on three levels of analysis – colonial, imperial, and international – to reveal governments' priorities and illuminates the nature of political relationships in the same way that studying meanings of and approaches to disease has served as a sampling device through which to understand cultures and societies.[13]

The historical record provides an opportunity to understand how a certain government arrived at a particular set of views about public health and reveals the values, norms, and assumptions that underlay the development of health policy.[14] It shows how political, economic, social, and cultural forces influenced the form and content of policy – reflecting government practices and priorities or combinations of conflict, negotiation, and compromise. It also highlights the forces that were often outside of public or policymakers' awareness and provides a sense of the options and alternatives that were possible.[15]

Although scholarship in the past 25 years has shifted the focus from teleological accounts of tropical medicine's conquest of disease to a more nuanced analysis of colonial medicine, an important lacuna remains in the current historiography. Scholars have not adequately addressed the role of global events and international treaties and organizations in the formation of public health policy in India.[16] Historians have shown that metropolitan and colonial governments necessarily considered international demands and pressures and the realities of colonial trade and production but have tended to neglect the connection between the development of public health policies in colonial India and the elaboration of an international health arena, with the exception of discussing the role of international sanitary conventions in relation to quarantine and the regulation of the *hajj* pilgrimage.[17] Anglo-Indian and Indian officials and scientists actively participated in interregional and transnational networks and discussions, and these connections were critical to shaping policy choices. The development of an interregional public health arena was closely linked to, but also distinct from, international health organizations.[18] However, tension existed between the growing involvement of colonial states in the international health arena and the lack or inefficacy of public health interventions at the colonial, national, and international levels to promote the health of subjects.

The majority of the scholarship on medicine or public health in the British Empire has focused on understanding the roles of medicine and public health measures in the colonizing process, as expressions or enablers of colonial power, as windows on colonial society, in nationalist appropriations, as tools for social control, in constituting the modern postcolonial nation, or in demarcating and constructing racial, cultural, and class divides.[19] What needs to be studied in conjunction with these roles are the reasons for which medical and health policies took a particular shape, which then allowed medicine and public health to be utilized for colonial, nationalist, or individual purposes. Epidemic outbreaks undoubtedly created disorderly and threatening situations, which enabled colonial governments to take actions against disease threats, thus reinforcing colonial rule.[20] But how did these colonial powers decide which actions to take?

Rather than view the colonial state's public health measures as primarily 'tools of empire', it is necessary to understand the motivations and influences affecting the colonial state's decisions.[21] This approach allows a more nuanced understanding of why the central government approved or avoided certain public health policies. Similarly, wholesale indictments of imperialism do not accomplish much in terms of understanding the role that imperialism may have played in both facilitating and hindering policy-making in India. Tension and contradiction were inherent to the colonial state's 'operative mode'.[22]

To achieve a more comprehensive understanding of policy-making, this study looks outside metropolitan-colonial circuits of knowledge and communication and employs a web-based model of analysis instead of the traditional hub-and-spoke view of imperialism to transcend center–periphery, domestic–foreign, and West–East oppositions. Moving beyond the borders of empires or nation-states is critical in understanding the interconnectedness of various parts of the world and in explaining the mechanisms and institutions shaping the movement of capital, people, and culture. In global histories, it is important to recognize that there were alternate conceptions of regions, which also served as zones of exchange, interchange, circulation, and transmission.[23] A recognition of the existence of interregional areas that were situated between a world system and a geopolitical region helps avoid a too macro or micro analytical approach to international history.[24] This book moves between an examination of decision-making processes and actors within and among states and decentered or local studies that focus on the political, social, and intellectual contexts within countries and that view international relations as a reflection of domestic conditions and ideologies.[25] The history of disease shows that epidemics emanated from diverse parts of the world and that dichotomies of West–East, metropole–colony were often irrelevant in developing strategies to prevent and contain outbreaks. The external global and internal local were in constant dialogue and conflict in government approaches to epidemic disease.[26] Epidemiological boundaries, like oceanic connections, follow the general contours of economic and social interactions.[27] Following those contours can avoid an artificial demarcation of geopolitical boundaries in analyzing the development of infectious disease policy and can demonstrate when and where connections and interactions occurred spatially and temporally and how those connections related to perceptions of risk in the context of infectious disease policy.

## Risk in the context of disease

The identification and conceptualization of risk depend on frames of reference and, of course, on who is doing the framing; it can include many different qualitative and quantitative factors and measures based on the frame of reference chosen.[28] Broadly defined, risk is the potential for suffering harm or incurring unwanted, negative consequences from a hazard, which could be substance, action, or event. Risk identification leads to risk characterization to inform decisions and address the needs and interests of decision makers and other relevant stakeholders. Characterization considers divergent views and estimates of risk; assumptions leading to risk estimates; the economic, social, ecological, and ethical outcomes of a situation; and how those outcomes relate to particular and entire populations. Risk assessment and analysis aim to identify, characterize, measure, and evaluate the nature and magnitude of risk and the outcomes resulting from different hazards. Addressing a risk situation, or risk management, involves understanding the relevant losses, harms, or consequences of the risk event to the interested and affected parties.[29] It includes the use of information on resources, response options, and values (social, economic, and political) to determine courses of action to minimize or eliminate risk.

Risk perception influences all stages of the risk assessment and management process. Individual risk perceptions vary depending on the likelihood and knowledge of adverse effects from a hazard, who is affected, the emotional fear of the effects, whether the individual has chosen to accept risk, and the personal effect of the hazard.[30] Once a risk is perceived, a decision must be made with the ensuing consequences, even if the decision is to do nothing.[31]

The growth in the literature on risk over the past 30 years provides many insights into risk in the context of public health policies. Sociological analyses of risk perception are informative in establishing a general framework within which to arrive at a political understanding of how states perceive risk. Scholarship has moved from focusing on risk and uncertainty as matters of science to recognizing that how societies and individuals prioritize risks, determine the extent of those risks that deserve attention, and decide what action should be taken are social processes determined largely by values and cultural and social preferences.[32] Research suggests that each culture or organization that has a set of shared values and supporting institutions is biased by those values toward highlighting certain risks and downplaying others.[33]

The governing structures within the British Empire, London, and India operated according to certain priorities and assumptions, which influenced policymakers' conceptions of risk. To understand how government policies unfolded, it is necessary to look at how these governments and relevant organizations, each with a particular political ethos, perceived and responded to opportunities created by policy-relevant events, such as the spread of epidemic disease.[34]

On the other hand, there are important gaps in the literature on risk, both generally and specifically in terms of public health. Research on risk and uncertainty has concentrated primarily on how people and groups perceive and understand risk. Focusing on individual and societal risk perception has had two consequences: neglect of the organizational and institutional contexts that have an important role in shaping those perceptions and little study of the formation of organizational and institutional perceptions of risk.[35] More research is needed to develop a theoretical perspective on the role of political power, governments, organizational agendas, institutional interests, and economic priorities in the formation of both individual and organizational risk perceptions, risk assessment, and the allocation of risk. The literature lacks detailed analyses that place the formation of risk perceptions and the production of knowledge about risks in broader political and economic contexts and that recognize the role of multiple and varied stakeholders and political and economic pressures in shaping understandings of risk.[36]

The literature proves even less helpful in addressing state or organizational perceptions of risk in the realm of infectious disease. Sociologists, such as Ulrich Beck, Niklas Luhmann, and Anthony Giddens, have focused on 'high-consequence' risks resulting from scientific and technological advances of the modern age and derived largely from the globalized character of modern social systems.[37] Beck's and Luhmann's analyses of risk focus almost exclusively on those risks created by or associated with technology in industrialized societies but do not address the diverse types of disasters and risks encountered today and largely ignore natural hazards, such as disease, famines, and natural disasters. Their theories do not encompass epidemic diseases and their spread, which was aided by globalization in both the early modern and modern eras. Beck discusses the politics of risk and the effects of manufactured risk on political organization, but not political risk itself, and overlooks the fluidity of risk as a concept, especially as perceived across different geographies.[38]

The existing scholarship on risk discourse in public health has tended to fall roughly into two domains: (1) risk as an externally imposed health threat to populations in which the individual has little control over the external or environmental hazard, or (2) internally imposed risks resulting from individual lifestyle choices where self-control can mitigate the risk. Epidemiology addresses risk in terms of the construction of risk factors and of categories of at-risk individuals or in the development of social responsibility to reduce risk at the population level by attention to individual health and lifestyle.[39]

The historiography on risk in relation to public health has had little to say as well on state perceptions of risk. By examining the place of man's physical well-being in doctrines of political economy, historians have indirectly arrived at an analysis of how state conceptions of risk informed public health investigations and hygienist doctrines.[40] During the nineteenth century, European hygienists and medical statisticians made the first attempts, based on the economic value of man, to calculate financial losses due to disease and to understand the implications of that loss to the state.[41] The position of European states on a disease's 'geoepidemiological trajectory' influenced their understandings of the imminence of a disease threat and affected state public health policies. Different nations construed disease and poor health as threats to the military, the individual, the collective, or the state.[42] Nineteenth- and twentieth-century France, Britain, the US, and Canada, with their particular political and cultural ideologies, were keenly aware of threats to their political and social orders and analyzed the potential impact of disease with an eye to those threats. Those analyses shaped decisions about whether and how health threats were addressed.[43] States recognized that death due to disease reduced the labor force and military capability and weakened the bonds of civil society and social structures. As such, disease could compromise all four dimensions of state capacity: institutional, technical, administrative, and political.[44] However, in many sociohistorical analyses, risk is understood as primarily internal. Only recently have scholars begun to analyze the implications of infectious disease to a state's economic and political relations with other nations.[45]

Most of the historiography on risk and public health has focused on risk at the level of the individual. Historians have analyzed how government public health policies have addressed individuals with different risk profiles, such as prostitutes, the military, or urban dwellers. They have also examined individual perceptions of vulnerability to disease as a consequence of industrialization and urbanization or of individual constitution, morality, behaviors, and other risk factors construed as such through prevailing theories about disease causation.[46] The epidemiologic shift from infectious to chronic diseases that occurred in the twentieth century encouraged the identification of social, environmental, and behavioral risks, such as cigarette smoking, that were statistically associated with chronic disease.[47]

What is lacking is an analysis of governmental approaches to address various types of risk to the state in the context of disease. How did states view and deal with uncertainty and the potential political and economic ramifications of disease spread and actions taken in response to looming epidemics?[48] To some extent, the recent literature on biosecurity and preparedness is yielding insight into how governments, particularly the US, approach preparing for the possibility of another pandemic.[49] One is more likely to encounter discussions of risk in this context in intelligence reports and in risk assessments for potential pandemic diseases. National security analysts label infectious disease as a national security threat when it compromises the territorial integrity and sovereignty of a state. Threat assessments take into account a disease's transmissibility, lethality and virulence, the potential to cause economic damage, and the fear it engenders in governments and populations. The framing of infectious disease as a national security threat, rather than as a 'human security' threat, structures the nature of government responses, the allocation of resources, and the parts of the bureaucracy involved in the response.[50] One study has argued that perceptions are shaped by countries' confidence in the affected country's assessment of the risk of disease transmission beyond its borders and in the transparency of communications from the affected country. International confidence in a country's self-assessment of disease risk depends on several factors in the affected country: effective disease surveillance; transparency in disease reporting; effective containment of outbreaks; strength of public infrastructure; maintenance of public trust in government measures; media access to information; and timely, accurate reporting to multilateral organizations.[51] Scholars have also suggested a linkage between state capacity and state security, arguing that those states with strong capacity will not view epidemics as threatening because they have more preparedness and response resources.[52] However, history shows that the ability of a state to address an epidemic does not necessarily correlate with its perception of the level of risk posed by a disease. We still have yet to conduct in-depth analyses from both historical and contemporary perspectives on the considerations, assumptions, and pressures that lead to preparedness policies and measures to address the threat of epidemic disease spread. We need a theoretical perspective on how qualitative understandings of risk are formulated by states and used in the development and implementation of policies and responses.

State perceptions of risk must be integral factors in any analysis of infectious disease policy. The 2009 H1N1 influenza pandemic, originally referred to as the swine flu, demonstrated all too clearly the importance of these perceptions. As news of the first H1N1 outbreaks erupted in the media, different countries began formulating and implementing policies that ranged from the rational (closing schools where cases had appeared in the US) to the irrational (orders to cull hundreds of thousands of pigs in Egypt). The spectrum of responses suggests that different countries had disparate perceptions of the risk that H1N1 posed to their health, economy, and political and social orders. Why do governments perceive risk differently, and what are the different factors that influence their responses?

This book represents a move to answer some of these questions. In looking at infectious disease policies historically, it aims to understand how government perceptions of political, economic, and epidemiological risk affected policy development to prevent and control epidemic disease. To do so, it analyzes perceptions of risk directly connected to the health consequences of disease and perceptions of risk related to the economic and political consequences of a response or set of responses to disease. The book then examines how those perceptions influenced policy decisions to manage those risks and the spread of disease.

In British India, government officials had primarily qualitative understandings of risk, with some exceptions made to quantify the economic cost of malaria and disease generally.[53] They viewed risk from disease in relation to the colonial economy, political relationships, and the health of the population. Economic risk encompassed negative consequences for international trade, domestic economies, labor productivity and supply, and transportation networks. Indian and imperial trade needed to continue with the least disruption possible, which meant preventing European trade embargoes on Indian goods. Maintaining the health of laborers, particularly in the labor-intensive tea, jute, and mining industries, was also critical to the viability of Indian production and to the greater British Empire.

Economic risk, especially considering the amount of Indian foreign trade, could not be easily disentangled from political risk. The handling of disease outbreaks had the potential to increase tensions in local, national, and international political relationships. The globalization of health issues and disease control threatened national sovereignty in each state's ability to decide what was best for itself, and, in turn, assertions of national sovereignty and imperial politics hindered disease control internationally.[54] The position of British India as a source of epidemic disease raised questions about the relationship between national jurisdiction and self-interest and a nation's responsibility to a larger international community. Britain and India resented disruptions to their trade and infringements on national and imperial sovereignties; however, they both realized that they had to accept limitations on their sovereignties to prevent political ostracism and foreign governments from imposing severe trade restrictions.[55]

In addition to addressing potential conflict in the international arena, the Government of India also had to hedge against jeopardizing relations with different constituencies within India, particularly when provincial priorities and local demands were at odds with the central government's objectives. In the post-World War I years, the emergence of an all-India mass nationalism changed the political context and led to a greater demand for political policies to reflect Indian rather than colonial interests. The Government of India also became more cautious about instituting certain disease control measures, such as restrictions on pilgrimage to the Hedjaz, with the increasing participation of Muslims in nationalist movements and with widespread Indian involvement in the Khilafat movement to protect the Ottoman ruler's position.[56] The central government was in a delicate and precarious position: it needed to avoid offending the religious sensibilities of its subjects by restricting participation in religious pilgrimages and festivals, but it also needed to protect its economic interests and foreign relations.

This framing of the economic and political consequences of disease spread was predicated on the existence, or at least perception, of a health threat. Epidemiological security was necessarily a critical consideration of public health policy, even if it was frequently viewed within conceptions of political and economic risk. In current models of risk, mathematical calculations are based on the assumption that absolute safety cannot be achieved. In effect, this makes security as a counterconcept to risk an empty and misleading concept and the risk/security opposition asymmetric.[57] Security 'acts only as a reflexive concept with the function of elucidating the contingent nature of the states covered by the concept of risk'.[58] In British, Anglo-Indian, and European qualitative understandings of risk from disease spread, security could not be a valid counterconcept either. The absence of a diametrical opposite to risk meant that measures to prevent the spread of disease were critical. Prevention was both a function of risk perception and a result of risk assessment; it meant reducing the probability of or mitigating the extent of future harm.[59]

Perceptions of the public health danger posed by disease resulted from understandings of disease ecologies and assessments of the impact of actual or potential epidemic crises on particular populations. Understandings of infectious diseases were not static, and the policy relevance of different diseases at the international, imperial, and colonial levels shifted over time as scientists advanced etiological knowledge and as disease morbidity and mortality rates changed. Laboratory and epidemiological research in India and abroad enhanced understandings of the biological realities of disease and influenced understandings of epidemiological risk. Medical research and the development of new medical technologies altered the risk profile of diseases and also changed the health policy choices available to governments – either shifting their emphasis or narrowing public health's purview. It also affected the types of policies that the international community could insist on and that the colonial state could implement.[60]

Not all diseases in India were subject to the same domestic or international scrutiny, however. Certain infectious diseases were more policy relevant and became foci for debates because of reasons of economics, visibility, infectivity, mortality rate, endemicity, and epidemicity. Plague and cholera in particular terrified Europe and claimed hundreds of thousands of Indian lives annually, menacing health officials on both sides of the Red Sea. Because of the high mortality and disruption to daily life and international trade and communications caused by these diseases, not to mention their gruesome manifestations, European states were quick to respond to epidemic outbreaks in India with the imposition of quarantines and embargoes. Even well into the 1920s, Europe and international health authorities remained most concerned about plague and cholera incidence in India, despite the immense toll that malaria took on Indian lives and the worrisome possibility of yellow fever spreading to Asia.[61] If the metropolitan and colonial governments wished to maintain the profitability of the jewel in their crown, public health policies needed to be enacted and quickly.

Malaria, on the other hand, debilitated millions, threatened the efficiency of India's labor supply, and resulted in substantial economic loss, but aside from the research conducted in India, the incidence of malaria in India itself did not garner much international attention. Adherents to the philosophy of the London School of Tropical Medicine, of which Patrick Manson was the most prominent, believed that since bacteria lived in the body and were rarely under the influence of the external climate, they could be transmitted in any climate under suitable conditions, accounting for European vigilance over India's cholera and plague outbreaks. However, the malaria-causing parasite, which was spread by several species of *Anopheles* mosquitoes, was climatically localized.[62] Furthermore, malaria, an infectious parasitic disease, was already prevalent in parts of Europe. While Europe paid little attention to malaria in India, from the Indian perspective, this endemic, hyperendemic, and sometimes epidemic disease should have been India's top public health priority. Malaria had the highest morbidity and mortality rates and economic costs among all diseases in India.

Yellow fever presented a different problem altogether. Unlike plague, cholera, and malaria, it was not present in India but threatened to spread, first from Latin America and later from Africa to India's coasts. In contrast to plague and cholera, yellow fever reversed the direction of border control; the Government of India had to focus on keeping disease out and on regulating endemic areas on other continents. India possessed three important factors, which created a climate favorable to yellow fever: several species of mosquitoes capable of transmitting the virus, a completely non-immune population, and a potentially large animal reservoir. Only the virus itself was lacking. While yellow fever in Africa was thought to pose a risk to India's population, public health experts realized that the disease did not seriously threaten Europe.[63] Yellow fever was largely confined to tropical and subtropical regions, and Europe lay outside of the latitudinal bands of endemic yellow fever. The mosquito vector, known then as *Stegomyia fasciata*, was not naturally found in Europe either. Thus, yellow fever, unlike plague and cholera, was a disease of European colonies, not of Europe itself – a distinction which critically affected yellow fever's risk profile.

The Government of India's infectious disease policies regarding plague, cholera, malaria, and yellow fever provide a fascinating lens through which to understand the dynamics of policy-making. Like many current governments in developing countries, it had to confront a daunting array of infectious diseases with constrained resources – financial, logistical, and medical – and in unique political, religious, and cultural contexts. The vast majority of India's 300 million people lived in rural areas, was malnourished or undernourished, lacked clean water, and did not have access to government dispensaries and hospitals. The physical environment and sanitary conditions in India aided the spread of plague and cholera and provided ideal breeding grounds for malaria-carrying mosquitoes. India's vulnerability to epidemics resulted from the coexistence of attempts to modernize infrastructure with inattention to persistent underdevelopment.[64] Colonial development policies to construct modern rail networks had led to ecological degradation, which created ideal breeding grounds for malaria-carrying mosquitoes. A widespread grain trade and high human mobility further helped disseminate disease, while poverty, undernourishment, impaired immunities, insanitation, and exposure to infection made Indians more susceptible.[65] In India in the early 1920s, the average number of deaths annually from preventable disease was estimated at five to six million.

The average number of working days lost per person from preventable disease was between two to three weeks; the loss of efficiency per average person from preventable malnutrition and disease was at least 20 percent; and the percentage of infants reaching a wage earning age was only 50 percent. While the colonial government may have focused on the loss of productivity, this measurement did not begin to capture the full spectrum of social and psychological consequences and economic hardship that disease visited upon millions of individuals and families. Preventable disease cost India immensely.[66]

## Public health in a colonial world

As a colonial government within the larger British Empire, the Government of India necessarily operated within the limits imposed by colonialist ideologies. Because of the nature of empire, there were economic and political constraints on policy, but health officials' negative attitudes toward Indian customs and lifestyles also often limited the scope of disease prevention and control. The colonialist view that Indians were resistant to change provided a convenient excuse for lack of political will to implement sanitary improvements. Yet, by the 1890s, disease control among the civilian population had become an economic and political necessity even for this colonial state, primarily because of changes in the international regulation of disease that occurred in the late nineteenth century.

The 1890s saw greater progress toward international health cooperation and the ratification of international sanitary conventions, which were aided by developments in bacteriology, parasitology, immunology, increased field studies in the colonized tropics, and a targeting of specific diseases like malaria and yellow fever for research, in part to address the diseases affecting colonized labor. Significant advances in etiological knowledge were accompanied by greater acceptance of reductionist notions of disease specificity and causation.[67] Health officials and scientists developed new approaches to addressing disease in the tropics; broad environmental and sanitary improvements were increasingly displaced by the targeting of specific diseases in and around individuals, disinfection to destroy pathogens, and isolation to break cycles of vector transmission.[68] These measures were not new, but the rationale behind them had changed in accordance with new scientific knowledge.

These developments in medical science facilitated international consensus on disease control. This contrasted to the period prior to the

1890s when forces restricting the outcomes and the scope of any result-ing international law had counteracted forces compelling states toward cooperation on disease control.[69] Britain itself had acted as a major obstacle to cooperation, firmly resisting European attempts to control cholera in British India as a form of 'continental despotism'. For most of the 1800s, the Government of India had contented itself with focusing its sanitary efforts on military cantonments and areas of European habi-tation, but that compartmentalized view of public health would end definitively in 1892, when Europe, Britain, and India delivered the first ratified International Sanitary Convention in an attempt to streamline quarantine practices and prevent the spread of cholera from India to Europe. By the 1890s, Europe, through the use of political pressure and economic sanctions, had forced the Government of India to recognize its obligation to a larger community of states and to accept responsibil-ity for managing disease in its territories. Disease control in India was necessary to secure the public's health in Europe.[70]

Policy-making during late colonialism continued to reflect this ten-sion between the Government of India's priorities (or lack of them) regarding disease control and the imperatives of international health. Examining this tension reveals both the conditions that led to a perceived need to make policy and the context in which policy was made.

Today, we take for granted the idea that one of the primary respon-sibilities of a modern nation-state involves protecting the health of its population – a view that has gradually evolved from a mercantilist phi-losophy dating back to fifteenth-century Europe, which placed the social and economic well-being, and thus the health, of the population in the service of the state. Public health is generally considered to encompass any state or state-sanctioned measure or action to protect or promote the health or well-being of a group of people, a society, a nation, or a group of nations. The control of communicable diseases and the reduction of infant mortality were seen historically as measures vital to European mercantilist policy. Therefore, public health has usually been depicted as the joint product of the growth of democratic and bureau-cratic states and the rise of epidemiology and later bacteriology.[71]

In the context of empire, these conceptions of public health need modification. The colonial state occupied a particular position in that it ruled over subjects, not citizens. Its establishment was based on a superiority of force; its existence justified through the preservation and extension of imperial power.[72] Enlightenment philosophies of democratic citizenship and nineteenth-century emerging conceptions of a right to health did not apply to British India, where the paternalism of a centralized state clashed with liberalism's assumption of the inequality of power in empire.[73]

Ideologies of empire emanating from Britain changed over time and affected the types of health policies the Government of India could or wanted to implement. The British failed to resolve conflicting ideologies about the applicability of liberal principles in empire. One camp held that liberal principles were universal; another maintained that the principles only applied to Western cultures; and the third believed that training and education could overcome cultural and historical differences to make liberal principles applicable. All three ideologies coexisted in the colonial state, and they could do so because they all rested on a colonial rule of difference. Supporters of empire justified 'unliberal' solutions abroad through the rhetoric of development and progress, even as the realization of that better future was perpetually deferred.[74] The conflict between liberal ideologies and the exigencies of empire and the inability to resolve tensions within a liberal discourse resulted in contradictory political, social, and cultural policies.

Since implied contractual obligations between states and their subjects did not apply in the same way throughout Britain's colonized tropics, why did public health policies even exist in colonial India? Initially the practice of Western medicine and the implementation of sanitary policies stemmed from a need to protect the military, European officials and settlers, and laborers in India.[75] The colonial state focused on protecting the health of the military and Europeans, with scant regard for the well-being of its Indian subjects.[76] Medical services provided by the British were inadequate to cope with India's health problems and provided little medical aid to the masses.[77] To its credit, the colonial government in India did create medical infrastructure in urban areas, but this did little to meet the necessities of the Indian population and only partly compensated for the health problems created by British land and labor policies. Sanitary reforms remained tied to a colonial imperative of safeguarding the health of military camps, European enclaves, and cities. Rural Indians suffered doubly from the government's lack of patronage for indigenous medical systems and from the withholding of Western medicine.[78]

By the late nineteenth century, Western medicine in India had outgrown its enclavist origins in the military and in prisons to encompass the urban Indian body as a site for constructing British authority, control, and legitimacy. The imposition of Western medicine on Indian society formed a critical part of the larger colonial project of hegemony, coercion, and control over Indian lives.[79] Indian health, sanitation, productivity, habits, and habitations became 'objects of governance' in the colonial state's quest for knowledge and regulation. Mortality and morbidity statistics became a means for objectifying the indigenous population.[80]

Health policies fulfilled multiple political, economic, and social objectives within the British Empire, but it is important not to lose sight of the fact that from the mid-nineteenth century, health policies served a very real purpose in preventing and controlling outbreaks of epidemic diseases such as cholera and plague and in protecting the overall health of the colonizers and certain sectors of the colonized. Politicians sensed that Britain in the 1890s was at her weakest economically and politically and was militarily overstretched and diplomatically isolated. Imperialists reacted with calls for national efficiency in labor, capital, intelligence, and reproduction, as they argued for the expansion of empire for its own sake and not just for economic or strategic ends.[81] The renewed imperialist fervor of the 1880s and 1890s brought to Britain, and to a lesser extent to India, a new focus on the health of the population and hence on nutrition, working conditions, and maternal and child care.[82] The dual forces of national interest and capitalist priorities led the Foreign and Colonial Offices to push British commercial and financial interests in colonies in Asia and Africa and concomitantly to lend material and political support to tropical disease research through the establishment of the schools of tropical medicine in Liverpool and London in 1898 and 1899, respectively.[83] In the 1890s and through to the 1930s, the capacity for commercial production and trade remained intimately tied to the state of disease and health in the colonies. From the late nineteenth century, policies to control epidemic diseases became necessary to protect colonial trade and to prevent the economic isolation of Britain and India.

This is not to say that trade and imperial strength were the only motivating factors in disease prevention and control. There was a sense that the colonial government could not wholly abdicate responsibility for the health of its Indian subjects. Central and provincial governments established dispensaries throughout India and built hospitals in major towns. However, the rhetoric of empire permeated government perceptions of risk and colored health policy discussions.

Empire is central to understanding public health policy development in colonial India, but it is critical not to view policy only though the lens of empire. This would overlook the international context in which policies in colonial India were discussed and elaborated. In examining infectious disease policy in India, Europe and its history cannot be merely 'silent referents'.[84] European states were the originators of international public health diplomacy, within which India was also a central protagonist, and India and Europe were inextricably bound through trade, the production of scientific research and knowledge, and patterns of communication, which created microbial exchanges. Sanitary regulation of not only India but greater Asia was a strategic decision that sought to protect commerce, achieve regional political gains, and balance expansionary aims of empire with imperatives of European diplomacy.[85] The notion that India had a responsibility to the world to prevent and control its diseases because of its trading position and because of its disease ecology gained greater currency within and without India. Although India was a colonial state, it had a diplomatic presence and vote in the international public health arena. The Government of India sent delegates to the international sanitary conferences, and, since India was a member of the League of Nations, health officials from India also served on committees in the League's Health Organisation.[86] As a result, it was in the peculiar position of being simultaneously subject to regulation by an international public health community and by an imperial center. Between 1892 and 1940, the Government of India became gradually more independent in its policy-making, but it always had to keep in mind the views of the British government and of the international community. The juxtaposition of Europe and India in the realm of public health policy illustrates where and how their histories diverged and converged and highlights instances of epidemiological, economic, political, and cultural interpenetration.[87] This book, thus, elucidates not only the workings of the colonial state but also the interactions of domestic and international public health authorities over time and as they related to policy discussions in India. It is at its core an international narrative that highlights the role of Europe when and where appropriate, but also provides a central, and by no means subordinate, place for India in that narrative.

Policy resulted from a nexus of interactions and relationships among these several layers of political and public health authority – Anglo-Indian, Indian, British, and international – and was a function of the rationalities and strategies of different, contending stakeholders, whose values, assumptions, and ideas created a framework for what was not only possible but desirable.[88] Policy was formulated within a multilayered structure that shifted throughout this time period. Stakeholders operated within and between international, imperial, and colonial political and economic frameworks and agendas.[89]

What 'international' encompassed politically altered throughout the course of the nineteenth and twentieth centuries. In the 1890s, the international community, as far as public health in India was concerned, constituted Britain, European states, the Ottoman Porte, and the Constantinople and Egyptian Sanitary Councils. This community gradually expanded to include the larger British Empire, the Americas, the Far East, and European colonies in Africa. Participants in international public health discussions expanded throughout the period from 1892 to 1940, even if not all of these participants sat at the diplomatic table. Within this broader international arena also emerged interregional forums such as the Pan-American Sanitary Bureau and the Far Eastern Association of Tropical Medicine, with representatives for India participating in the latter. From the early 1900s, intergovernmental organizations began entering the international public health arena in the form of the Office International d'Hygiène Publique (OIHP), established in Paris in 1907, and the Health Organisation of the League of Nations, formed in 1923.[90] These organizations primarily served as forums for international health discussions, helped prioritize international health goals, advised governments on public health matters, and coordinated actions among states related to the prevention and control of disease.

Operating parallel to and often within this international diplomatic structure was an imperial one that consisted of three main centers: London, where the Secretary of State for India sat and answered to Cabinet and Parliament; Calcutta and later Delhi, where the Governor-General or Viceroy ruled over India with an appointed council and legislative assembly; and the provincial capitals, where governors had their counterparts of the central machinery. Parliament was the ultimate authority and metropolitan priorities generally took precedence over the political and financial needs of the colonies; however, most policy decisions with regard to India were reached through compromises made by officials at these three centers, whose priorities differed based on local situations, pressures, and audiences.[91] All of these layers were affected by infectious disease diplomacy at the international level. In the absence of international policy, the metropolis and the colony had more freedom to structure health policy, but conflict among the three centers could still arise, particularly when what was beneficial politically, economically, or sanitarily for India held disadvantages for other parts of the empire.

Despite being enmeshed in several levels of authority, the Government of India, in matters concerning public health and disease control, largely operated independently of the metropole, meaning Parliament, the Cabinet, and the India Office. However, a distinction between the colonial domain and the metropolitan one could not always be maintained; the British government intervened when disease prevention and control threatened trade, other parts of the empire, or Britain's political relationships within Europe.[92] When these situations occurred, power struggles between the British and Indian governments could result, which brings into question the idea of a single 'official mind'.[93] If, on the other hand, British and Indian government officials concurred in how to address epidemic disease control, this agreement strengthened their negotiating position in the international public health arena.

Within India, the central government operated within a pseudo-federalist model, in that power was shared, albeit unequally, between the center and the provinces. The British government in India, with its goal to protect the interests of British industrial and commercial capital, concentrated real power at the center. Divisions of power among various levels of government and colonial laws, bureaucracy, and administration of justice reflected the slippery space between conquest and occupation that the British inhabited. The simultaneous collaboration with and displacement of Indians in government service resulted from the colonial state's hesitancy to choose decisively one form or the other.[94] As the political climate in India changed with the rise of nationalism, less critical responsibilities were increasingly relegated to provincial and local authorities in an effort to divert Indian attention away from the center. Provincial governments were generally responsible for matters that lay outside of defense, foreign affairs, fiscal policy, and all-India concerns – areas over which the viceroy would retain tight control, even after the constitutional reforms of 1919 and 1935. Responsibility for public works and sanitation were off-loaded onto local Indian boards and municipalities, while budgets and administration remained under central surveillance, ensuring that the colonial center retained ultimate authority.[95] In concessions to Indian nationalist demands during the interwar period, the Government of India granted provinces greater autonomy and a wider jurisdiction, enfranchised Indians, and established provincial legislatures. Yet, it must be remembered that these constitutional changes moving Indians towards greater 'self-government' were made in the context of a British government determined to retain its empire.[96]

In health matters that concerned all of India or interstate relations, the Government of India remained the final authority, with the Sanitary Commissioner (later Public Health Commissioner) for the Government of India residing at the top of the public health bureaucracy. Provincial governments did, of course, assert their interests in formulating policy based on administrative capacities, disease incidence in their respective provinces, and the impact of health policies on their province's trade. The Bengal, Bombay, and Madras presidencies contained the largest trading ports in India, and since trade was intimately tied to public health policy, these three provincial governments were privileged in the central-provincial dynamics that shaped policy.[97]

These public health policy discussions were not limited to state or intergovernmental actors but drew the involvement of several groups of non-state actors as well. Because prevention of the spread of epidemic diseases outside greatly affected the flow of trade and the health of laborers, trade associations and mercantile organizations actively participated in policy debates with central and provincial governments. These powerful lobbies were at times able to change the course of health policy in order to support their commercial interests. Medical officials and scientists formed another influential group. Their knowledge and opinions were critical to deciding the types of policies the colonial state could and would implement. Advances in bacteriology and parasitology had broadened understandings of what constituted the environment of disease and had led to a more precise understanding of causal organisms, incubation periods, vectors, and the role of sanitation, enabling the control of communicable diseases on an unprecedented scale and in a more targeted manner.[98] The findings of scientists and colonial medical commissions guided policy debates and provided scientific bases for policy decisions.

Indians also influenced government policies on infectious disease, albeit in a limited manner.[99] At times, Indian sensibilities and reactions to policies were taken into consideration in determining which policies to implement and how. Informally, through resistance to offensive measures, Indians did affect the decision-making process, as seen most dramatically after violent protests during the 1897 plague outbreak prompted the assassination of the Plague Commissioner of Poona. Regarding sanitary regulation of the *hajj*, the colonial state sought the opinion of educated and elite Muslims as part of its efforts to court cooperation of the Indian elite. The development of the nationalist movement, particularly after World War I, altered the political context in which policy was made, and the central government took this changing climate into account. In the twentieth century, the Muslim elite, which had been consulted on pilgrimage policy, began to be regarded with some suspicion as a politics of pan-Islamism became more popular.[100] Throughout this period, Indians also critiqued government policies in newspapers and journals and pressured the Government of India to varying effect to attend to the religious and cultural sensibilities of its subjects.

What is remarkable is that Indian voices were hardly ever considered formally during policy discussions. Promotion of the overall health of Indians was only a minor consideration in the structuring of epidemic disease policies, resulting in a notable absence of the Indian perspective in the policy-making process. Furthermore, the Indian National Congress was more interested in reforming the structures of medicine than in addressing the determinants of health and disease. Prior to 1920, the Congress focused mainly on improving opportunities for Indians to enter and advance in the Indian Medical Service. After 1920, the Congress shifted its attention to deciding what type of medical system (Western or plural to include Ayurveda and Unani systems) it should support and as a corollary to this, to promoting medical registration of practitioners and the international recognition of Indian medical degrees. Outside of official forums, nationalists, and Mahatma Gandhi in particular, focused on improving Indian diets as related to physical, mental, and moral fortitude. They viewed nutrition in terms of the health of the future nation and linked dietary reform to India's prosperity.[101] By 1940, the National Health Sub-Committee had directed its efforts toward capacity-building. The sub-committee set a target number of health workers and beds for every 1000 people and recommended the absorption of Ayurvedic and Unani systems into the state health organization through the provision of scientific training. It also advocated a centralized public health agency, the administrative integration of curative and preventive functions, and self-sufficiency in the production of drugs and medical equipment and supplies.[102]

While nationalists focused on medical infrastructure and Indian lifestyles, colonial officials remained largely disinterested in obtaining Indian stakeholder input into policy debates. Thus government officials, scientists, medical doctors, and international organizations became the primary actors in policy development on epidemic disease, leaving Indians on the margins.

There is, of course, a story to be told about the effect of policies on Indian lives and on the relationship between the individual and the state. However, this is beyond the scope of this work, which seeks to understand how international and colonial perceptions of risk influenced policy-making.

## The disease approach

Much of the historiography on disease in India has assumed a monolithic colonial medicine and has applied arguments about the Government of India's health policies and motivations to a host of various diseases, thereby glossing over the particularities of place, time, and disease.[103] Alternatively, disease-specific studies have highlighted the dramatic infectious diseases, like cholera, plague, and leprosy, while giving less attention to the not as dramatic, but equally pernicious, infectious diseases like malaria, hookworm, and rabies and scant attention to the prevention of diseases that had yet to infect India. This study disaggregates government approaches toward four diseases to allow for the importance of distinct epidemiologies, varying depths of etiological knowledge, and differences in modes and efficacies of treatments and strategies for prevention.

Cholera, plague, malaria, and yellow fever presented three distinct risk scenarios: plague and cholera were both international and domestic threats; malaria was the most fatal disease in India and thus the most significant internal disease problem; yellow fever threatened to invade India but not Europe. These were also all diseases for which the state could take decisive measures at a community level (even though it also took action on an individual level).[104] As such, they can be utilized as prisms through which to analyze the Government of India's rationale behind its infectious disease policies, both short-term and long-term. An analysis of the circumstances leading to policy also reveals the obstacles that influenced policy directions. Why, at times, was there not a policy? Given certain perceptions of risk, why did the Government of India not make consistent policy decisions, and what obstacles impeded policy development? A longitudinal approach then shows how the type and degree of risk varied over time, while a case study approach bridges issues of causation and response and offers insight into policy continuities and discontinuities.[105]

Chapter 1 surveys the international context within which infectious disease policies were debated and structured in the 1890s. It examines how and why international public health treaties became a major factor in Indian public health during this period. It looks at the role of British India in the translation of an international public health consciousness into policy and examines the effect of that transition on public health policy in British India. The demands of international health security, imperialism, trade, and pilgrimage all coalesced to make India central to the early development of international health policy, which in turn influenced public health measures in India.

Chapters 2, 3, and 4 analyze each disease scenario. Chapter 2 on plague and cholera underscores how the importance of these diseases to both India and Europe affected policy decisions and highlights the Government of India's different approaches to epidemic versus endemic disease. Chapter 3 focuses on malaria and places a greater emphasis on colonial politics and economics in understanding why malaria policies took the form that they did. It also aims at a deeper analysis of the obstacles to policy development. Chapter 4 on yellow fever policy looks at the pre-emptive strike in a sense and aims to understand how a particular government addressed the issue of preventing the entry of a foreign disease. It examines how the Government of India with very constrained resources attempted to achieve domestic health security.

Chapter 5 brings together the case studies to arrive at a high-level understanding of why the Government of India made certain choices regarding four deadly infectious diseases. It highlights how the biological realities of different diseases interacted with political regimes, international communities, and economic, cultural, and social factors to influence the final form of infectious disease policy.

This study concludes with World War II because it was an important transition period for public health and colonialism for several reasons. First, medical resources and personnel were diverted from India to the war effort as colonial health policy shifted to address the imperatives of protecting soldiers' health.[106] Second, the scope and face of international public health efforts fundamentally changed as the war brought to light new biomedical and technological advances, such as penicillin and DDT. The altered international climate after the war's conclusion would also lead to the establishment of the World Health Organization, which both built upon the achievements of this earlier period in international health cooperation and marked a philosophical shift in thinking about health in the context of rights.[107] But perhaps most importantly for this study, India's financial relationship with Britain had significantly changed. The interwar decline in world demand for Indian goods and commodities affected India's place in the imperial economic system, and by 1939 the Indian economy had become less complementary to Britain's. World War II subverted the center–periphery relationship, turning India from Britain's debtor into its creditor, thus eliminating much of the economic rationale behind maintaining India as a colony.[108]

An analysis of this roughly 50-year time period provides an opportunity to compare policies for several diseases, since certain diseases garnered more or less of the spotlight at different times. On the other hand, covering a long span has the disadvantage of sacrificing some depth and continuity in the telling and makes it more difficult to focus on the roles of specific individuals and their political and health agendas. However, I hope to impart an understanding of how each disease presented a certain set of political, economic, and epidemiological risks, how perceptions of those risks changed over time, and how understandings of those risks and a constellation of factors combined to influence infectious disease policies in India.

# 1
# All Eyes on India

The mid-eighteenth century marked a turning point in Europe's epidemiological relationship to the Indian Ocean zone. European expansion of territorial empire around the Indian Ocean, advances in transport, and the intensification of trading all combined with the growth of urban and port cities, the movement of troops, pilgrimage, poverty, famine, insanitary conditions, labor migration, and ecological degradation to increase the circulation of disease.[1] Within the international order of nation-states and empires, disease piggy-backed on older forms and patterns of globalization.[2] It traveled west over land and in multiple directions via maritime trade and pilgrimage routes. Ports and cities acted as 'disease entrepots', receiving, incubating, and disseminating disease around the ocean and further afield.[3] Aided by these long-established and newer connections, cholera pandemics repeatedly swept over large parts of the world throughout the nineteenth century, and in 1896, plague once again resurfaced in epidemic form. The stony blue-gray faces of cholera victims caused by rapid and severe dehydration and the ulcerations and buboes indicative of plague provoked fear and panic across the world.

The eruption of cholera and later plague in India caused European states to fear the very real possibility that these 'Oriental' diseases would infect their lands. Although plague and cholera were not at all unique or limited to Asia, colonial discourse had reframed these diseases as scourges from the East, resulting from Asians' insanitary and decadent lifestyles and the debilitating effects of tropical climates on the body. Europe needed protection from these Asian invaders. European diplomats and health authorities pointed an accusatory finger at India, holding the Government of India responsible for preventing the spread of disease beyond India's borders.[4] As international trade grew, became more rapid, and intensified global connections, states and colonies reacted to epidemic disease outbreaks by imposing quarantines in the hope of preventing ingress of the disease into their territories.[5]

Although states relied on quarantinism to ward off epidemic disease, the unintended cost of quarantine was commercial chaos on the seas. Such disorder became increasingly unacceptable to European trading powers.[6] States realized that haphazard quarantine damaged trade more than it protected against disease. International trade and travel and the transnational nature of cholera and plague undermined the ability of Europe and the Ottoman Porte to prevent and control infectious diseases in their territories. Epidemic disease control measures would need to be formalized through congresses and conferences in order to meet the foreign and colonial trading priorities of European countries.[7]

States grasped the need for multilateral public health agreements to safeguard their populations from diseases originating in Asia, particularly India, considered then the 'hot-bed' of cholera.[8] The spread of epidemics had created a 'world risk society', in which states recognized that there were global problems that transcended national and local spheres and politics and that required a transnational framework, international institutions, and 'cosmopolitan parties' that could 'represent transnational interests transnationally', but also operate within the realm of national politics. Disease was a world problem of common concern that could not be addressed within a schema of national politics.[9] Without international cooperation, states realized that their efforts to control epidemic disease would be largely futile, prompting European states to convene the first International Sanitary Conference in 1851.

International efforts to protect Europe from Asiatic diseases were aided not only by commercial interests and the need to reform quarantine practices but also by changes in the European political climate and practice of diplomacy that followed the peace of 1815 and led European states in the mid-1800s to envision systematic and mutually beneficial sanitary control among nations.[10] This vision resulted in the International Sanitary Conferences, through which a nascent internationalism, in tension with European nationalism, became a significant part of public health policy in the latter half of the nineteenth century.[11]

The first six conferences held between 1851 and 1885 failed to yield ratified conventions, but they did reflect and support the development of an international public health consciousness and did formally internationalize public health.[12] The global spread of epidemics had created mutual dependencies and vulnerabilities between Europe and its colonies, and those interdependencies promoted interest in collaboration and gave rise to international health politics.[13] Early public health internationalism consisted of a combination of legal, political, and scientific interchanges and exchanges, which themselves resulted from growing economic internationalism.[14] Conferences and their respective conventions were framed within a 'Westphalian' system of international public health that was based on the principle of non-interference in the internal affairs of other states. In practice, European states selectively applied this Westphalian concept, infringing upon the sovereignty of non-European states within the Mediterranean region on grounds of pre-emption. Europeans perceived the Ottoman Porte as a weak state incapable of effective public health surveillance and sanitary control of its borders.[15] Collective action was necessary to compensate for the shortcomings of the Ottoman Porte and other 'eastern' states.

While the International Sanitary Conferences had succeeded in bringing disease prevention to the international stage in efforts to reconcile the freedom of communication, commerce, and trade with the protection of public health, states would not make concrete progress toward this goal for another 40 years.[16] In earlier conferences, the British government, Italy, and northern European states had opposed increased international quarantine regulation in an effort to protect their foreign and imperial trade.[17] Britain doubted quarantine's efficacy as a public health measure and consequently chose not to impose quarantine on its ships coming from ports infected with cholera or plague. British officials argued that they had successfully prevented cholera epidemics in Britain through selective medical inspection and without the use of quarantine.[18] By 1892, other European countries had also come to realize the futility of quarantining ships in preventing the spread of cholera into Europe, especially since cholera had traveled overland from Asia. As a result, they proposed relaxing quarantine controls, which proved crucial to obtaining Britain's and India's support for the international sanitary conventions.[19]

In the 1890s, an international health consciousness finally transformed into international policy. The lack of scientific consensus on disease etiology had impeded the ratification of any treaty before 1892. International agreement on preventive measures was facilitated by the elaboration of the etiology and epidemiology of cholera, plague, and then yellow fever and by a simultaneous state sanctioning of this new laboratory-based, scientific medicine.[20] A de-emphasis on quarantine as a primary preventive method and the gradual acceptance of Robert Koch's theory that cholera was caused by a bacillus permitted the liberalization of quarantine rules.[21] The four conferences between 1892 and 1897 resulted in ratified conventions, and marked a transition to the 'neoquarantinism' advocated by Britain and India – a practice that rejected old-fashioned quarantine, preferring medical inspection and surveillance of travelers, disinfection of people, ships, and merchandise, notification of disease to port authorities, and shorter periods of observation for ships. Neoquarantinism balanced the demands of public health and safety with trade and economic concerns.[22]

This updated form of quarantinism validated British and Anglo-Indian views on quarantine, while the recognition and formalization of the need for international control of disease proved a watershed moment for the development of India's epidemic disease policies. In the 1890s, epidemic outbreaks of cholera and plague in India directed Europe's attention to sanitary conditions and disease control measures in India and on ships sailing from India. The ratification of international sanitary conventions created a legal basis for Europe's scrutiny of India's response to disease outbreaks. The ratified conventions also obligated India to respond to cholera and plague outbreaks according to certain guidelines. Because Europe regarded India as the home of cholera and a reservoir of plague, the provisions of the international sanitary conventions would influence the form and substance that infectious disease policy would take in India for the next 40 years.

European delegates to the conferences held the disease-originating country responsible for preventing the spread of epidemics beyond its borders. This meant that India had a duty to contain disease. At the 1894 International Sanitary Conference, delegates convened to discuss how to prevent the spread of cholera via pilgrimage routes to the Holy Lands and subsequently to Europe. The French delegate, Henri Monod, argued that, '"Europe did not know cholera before India became a British possession. It is therefore principally the British Empire that has the responsibility of opposing its exportation."'[23] Since British ships were the primary transporter of pilgrims to Mecca, it would have been better, argued a prominent Viennese epidemiologist, if the British '"subordinated their petty love of gain to higher sanitary interests."'[24] Whether cholera had been imported by sea from India to the Hedjaz or to Europe remained a point of contention between British and other delegates at the conference; however, the pinpointing of the geographic origins of disease placed the onus of control on the originating country, which for cholera and, at times, also for plague, meant India.[25]

The cholera epidemics that invaded Europe from India in the 1880s and 1890s focused Europe's attention on India and attempts by Indian authorities to lessen the frequency of epidemics.[26] Cholera mortality in India had steadily increased, reaching a staggering 762,695 deaths during the 1892 epidemics. The 1892 outbreak in Hardwar (in current Uttar Pradesh) spread overland through Punjab, Afghanistan, Persia and, within five months, had hit Russia and continued moving west to Hamburg.[27] European delegates to the 1894 convention feared a repeat of 1892. They viewed these statistics with apprehension for their countries' health, while merchants in India and Britain worried what it could mean for their trade.[28] The Government of India thus came under significant international and mercantile pressure to abide by international health regulations. However, continued international disagreement over quarantine practices, the protection of British and Indian trade, India's subject status as a colony, and the assertion of India's unique disease ecology would not make the application of sanitary conventions to India's policies straightforward.

Despite the general view that Europe needed safeguarding from disease, consensus on how to accomplish this was not easily achieved. European views regarding quarantine practice differed significantly from those of Britain and India, both of whom strongly objected to quarantines and sanitary cordons because of fears of lost commerce. At the 1894 International Sanitary Conference, the British delegate claimed that 98 percent of shipping in the Persian Gulf was British, accounting for Britain's firm objection to tight maritime controls.[29] Britain preferred to rely on sanitary improvements at its home ports as the best prophylactic against cholera with only very limited recourse to quarantine; whereas, on the Continent, despite waning support, the belief in quarantine as a frontline measure persisted. The imposition of quarantine was further complicated by its connection to the control of shipping channels. Britain wanted to prevent damage to its commercial interests by avoiding rigorous enforcement of quarantine for ships passing through the Suez Canal. The Marquis of Salisbury accused the Continental Powers of treating the Suez Canal as an international domain, which European Powers could open or close at will, with the purported aim of preventing the spread of cholera to their countries. European states supported the idea of using the Suez as a controllable maritime gateway through which cholera could be stopped on its journey from India to Europe, particularly since not all European countries had the means or money to implement sanitary precautions at their borders as Britain had done.[30] Great Britain, on the other hand, 'contended that no restrictions are legitimate except such as the Khedive of Egypt, the lord of the soil, may find it necessary to interpose in order to protect those parts of his own dominions which are touched by the Suez Canal'. Because 77 percent of the tonnage passing through the canal was British, Britain had insisted that authority over shipping in the canal reside not with Continental powers but with the Khedive, over whom Britain wielded more influence.[31] The large amount of trade flowing between Britain and India necessitated the protection of British maritime commerce, which was linked to European quarantine practices, which in turn was influenced by the sanitary conditions of Indian port cities and ships. Thus Britain and India had a vested interest in convincing Europe that Indian ports and ships posed little threat to Europe.

## The Government of India mobilizes

The need to allay European anxieties while protecting British trade interests continually occupied the minds of delegates from Britain and India at the conferences. They had to devise sanitary measures for ships that would achieve both objectives. Britain negotiated with France to reach a compromise that all ships leaving India for the Red Sea would put into port for a sanitary inspection at Aden, which, conveniently, was under British control. No ships would be allowed to leave Aden without a certificate of health. If there were cases of cholera on board, the sick would be disembarked and the ship submitted to cleaning and disinfection.[32] Britain negotiated the compromise and port authorities in India implemented it on India's coasts. They carefully monitored the health of ships leaving Bombay, Calcutta, and Madras ports for foreign destinations.[33] In doing so, they hoped to reassure Europe that Indian ships did not pose a disease risk while expediting the passage of ships leaving India.

Europe had worried primarily about the spread of cholera from India via maritime traffic and pilgrimage, but, in 1896, cholera would temporarily exit the world's spotlight as international attention shifted to plague, upon the spread of the epidemic from Hong Kong to Bombay in late September. Not until after plague had erupted in India did Britain and the Continent realize the significance of the epidemic in China for health and trade globally. The Government of India, in contrast, continued to deny publically the existence of plague in Bombay for ten days after cases appeared in the city.[34] As countries on both sides of the Atlantic hastily imposed quarantine against ships sailing from India, the Government of India realized the futility of its strategy of denial.[35] As a further precaution against the contagion, many European states issued embargoes on goods exported from India. Italian and French governments prohibited the importation of hides from Calcutta due to plague in Bombay and Karachi, causing great inconvenience and financial loss to members of the Bengal Chamber of Commerce.[36] Gibraltar declared that it was not giving practique to vessels coming from India. (Practique was a government permission to conduct business at ports after ships had complied with the requisite sanitary and quarantine regulations.) Passengers and goods from India were not to be landed on Gibraltar soil at all. Malta imposed a 21-day quarantine against vessels from India. Portugal refused to admit vessels from British India, not even to quarantine. Germany, the United States, Peru, Russia, and South China, among other countries, imposed quarantine against all vessels from British India. Turkey imposed a 15-day quarantine and also prohibited the importation of animal products, while Antwerp and Spain banned the import of all textiles from India.[37]

The rapid international response to plague in Bombay created a dire situation for India and jeopardized its trade. In January 1897, the Secretary of State for India, Lord Hamilton, came under intense pressure from the Austro-Hungarian and Russian governments and London commercial interests to suspend all pilgrimage from India for the remainder of the year, since European states viewed pilgrimage to the Hedjaz as the most important mode of disseminating both plague and cholera. Cotton, tea, and hide merchants in India and Britain feared further international restrictions on trade if India did not impose a ban. India occupied a critical role in a multilateral trading system, and the suspension of foreign trade would have repercussions throughout the empire.[38] Other imperial powers, including France, had already suspended pilgrimage from its colonies with Muslim populations. Viceroy Elgin, however, was reluctant to put at risk Anglo-Muslim collaboration efforts in India, particularly after pro-Ottoman sentiment among Indian Muslims surged following the 1894–5 European campaign against Ottoman treatment of Christian Armenians.[39] Elgin argued that prohibiting pilgrimage would be seen by Muslims as interference in the practice of their religion and preferred to avoid the possibility of public outrage that might ensue following a ban.[40]

Britain and the Government of Bombay, on the other hand, wanted to ensure that India took adequate steps to contain plague, in order to pre-empt European retaliation against India at the upcoming 1897 International Sanitary Conference in Venice. In anticipation of the conference, the Government of Bombay tightened inspection of ships bound for ports in India and abroad and introduced railway medical inspection and observation and detention camps for passengers.

Bombay's Port Health Officer, Lieutentant-Colonel F. F. MacCartie, remarked that the main goal of these ship inspections was to prevent the spread of plague to other ports inside India, to ports outside India, and, '"above all, to Europe, with its incalculable effects on Indian commerce"'. No measures would be spared to obtain that end. Between February 1897 and May 1899, inspectors examined over 2,200,000 passengers and crew members in Bombay and refused travel privileges for more than 28,000 of those inspected.[41]

With Britain's Muslim subjects in mind, Elgin remained reluctant to suspend pilgrimage from India, but Hamilton would not risk Europe blackballing Britain at the Venice conference. Britain feared that continued international embargoes would not only close a critical market and source of raw goods but also would disturb Britain's multilateral balance of payments, for which Indian revenues were central.[42] Hamilton used his authority to force the Government of India to suspend pilgrimages to the Hedjaz and to prohibit Hindu religious festivals for the remainder of the year. The Government of India would need to do everything possible to stop pilgrims at the border and near their homes and to prevent large gatherings.[43]

Hamilton had made a wise strategic move. Foreign governments at the conference appreciated India's ban on pilgrimage, but the ban did not go far enough to appease European states completely.[44] India suffered from poor governance, living conditions, and public health infrastructure. As a result, unaffected Europe lacked confidence in India's ability to respond to complex health emergencies and was likely to implement countermeasures, such as embargoes and quarantines, to protect its European citizens.[45] Officials in India understood that international confidence in their anti-plague measures was critical to maintaining trade flows and positive economic and political relationships with Europe. The Government of India felt compelled by Europe, Britain, public health officials in India, and mercantile organizations to implement more drastic measures to contain plague. Local authorities in the Bombay presidency began searching homes for suspected plague victims, removing inhabitants from plague-infected dwellings, and destroying infected articles of clothing.[46] In Bombay, officials evacuated and disinfected infected homes and sent evacuees to special camps. Authorities monitored all passengers arriving by rail and recorded their names and addresses to track their whereabouts and health.[47]

The Government of India had hoped to avoid such drastic measures, which it knew would not be received well by Indians. It had sought to protect the religious and cultural sensibilities of its subjects, but the need to lift European trade restrictions on Indian exports and to reassure the international health community of the seriousness of the Government of India's efforts overrode domestic considerations. Debates among officials show that the Government of India was reluctant to impose house-to-house searches for plague cases or suspend religious pilgrimages and holy festivals, but they did impose such searches in response to international political and trade pressures.[48]

The combination of a ban on pilgrimage and medical inspection of rail and ship passengers and their effects achieved the desired result internationally. At the 1897 Venice conference, delegates expressed their approval of the thoroughness of the measures enacted in Bombay. In addition, foreign scientific commissions that had been dispatched to study plague in India reported favorably on the comprehensiveness of the inspection system. The Government of India's measures had convinced European officials that India no longer presented a plague threat. Officials responded by relaxing quarantine regulations throughout Europe against Indian ships.[49]

British delegates had attended the 1897 conference not merely to appease Europe and the Ottoman Porte at all costs but to avoid unnecessary encumbrances on the shipping trade and on the Government of India. Britain's Foreign Office sent instructions to Venice delegates that the interests of British ships, Muslim religious feelings, and the means at the disposal of the Government of India were to be remembered at conference negotiations. The Government of India should be careful about accepting and acting upon conclusions made by British delegates, since it would be impossible to repudiate them at a future time.[50] British delegates did their part by obtaining allowances for both Britain and India. Britain ratified the 1897 convention with some reservations.[51] Despite the threat and actual institution of trade embargoes to force Britain and the Government of India to control plague, Britain was still able to negotiate more favorable terms for Britain and India, since Britain's ratification was critical to the authority of the convention.

Efforts to win international approbation had domestic consequences, however. The Government of India's swift action, while reassuring to Europe, met with popular resistance in India. Radical and moderate nationalists, including Bal Gangadhar Tilak and Gopal Krishna Gokhale, decried the government's measures as tyrannical.[52] Some Indians were suspicious of the ideas and practices of Western medicine and of British intentions, while others deplored not the anti-plague measures themselves but the manner in which they were implemented.[53] The Gujurati newspaper, *Jam-e-Jamshed*, wrote that the Government of India should have 'evinced the same vigilance and solicitude for its Indian subjects which it display for the benefit of European nations and trade . . . It is only when the European nations are stirring to take stringent measures to prevent the spread of the infection to their countries that the Government of India has introduced a Bill to deal with the matter'.[54] The *Gujurati* criticized the 'indifference and callousness which His Excellency Lord Sandhurst's Government have shown in the actual execution of plague operations'.[55] Despite the Government of India's admonitions to attend to the sensibilities of Indians, local officials implemented measures with little regard for Indian social, cultural, and sexual norms.[56] Authorities disinfected buildings where plague cases had appeared or were suspected; burned bedding, clothes and infectable articles; removed tiles and partitions to allow light and air ventilation; moved the sick to hospitals if isolation was not possible or if patients were destitute or had been deserted; cremated the Hindu dead regardless of caste; burned infected huts; and medically inspected passengers leaving Bombay by train.[57] The transgression of customs, particularly in the inspection of women and indiscriminate segregation of the ill, led to widespread discontent and mistrust. The Indian press criticized these violations as well as the inadequate provisions at quarantine camps and the lack of compensation for destroyed property, which weighed heavily on the poor.

On the other hand, some Indian-owned newspapers accused the government of not going far enough. The Anglo-Gujurati paper *Kaiser-e-Hind* reported that 'never were the arms of Government so much paralysed, and never was a Government found so absolutely wanting in energy, as the inert local self-government of Lord Sandhurst and the nerveless Government of Lord Elgin'. The paper advocated the 'wholesale removal' of Bombay citizens until the disease had been brought under control.[58] At the same time, these newspapers implored the government to use more tact.

Popular protest against the government's methods continued to increase, however, and eventually culminated in the assassination of the Poona Plague Commissioner, Walter Rand. The assassination and widespread agitation finally caused the colonial state to amend its methods of disease control and to insist on greater consideration for the sentiments of Indians regarding the treatment of their persons, even while some officials viewed opposition to public health interventions as a sign of ignorance and superstition.[59]

The Government of India and provincial governments did not want to upset the delicate balance between public support for containment of the disease and outrage at the measures implemented to do so. The

Government of India did not countenance a relaxation of preventive and precautionary measures but recognized that enforcement of these measures had in some instances led to alarm and discontent.[60] As a compromise, the Government of India allowed local governments to relax plague regulations where public cooperation safely permitted or when popular tumult threatened, as was the case when rumors circulated about the government's supposedly sinister motives for implementing certain plague policies.[61] In September 1898 the Government of India issued amendments to its earlier guidelines for the conduct of plague operations by provincial governments. The Government of Bombay set up caste committees to manage the segregation of patients in hospitals according to caste and class.[62] The Government of India also ordered the elimination of corpse examination, removed the obligation to report illness or death from plague to the police, and permitted public and private hospitals and camps to be placed under the charge of indigenous medical practitioners. Those having received the anti-plague inoculation within the previous six months were exempted from segregation unless infected, but they were still liable to evacuation.[63]

While the drastic measures implemented in India had helped persuade delegates at the International Sanitary Conference that India was taking all necessary precautions, India had still not succeeded in containing the epidemic by 1898. The Government of India chose then to abolish these remedial measures. Detention of large numbers of people suspected to have the disease had essentially closed channels of business and was therefore inexpedient in the long term. In place of detention camps, a surveillance system was introduced in October 1898, whereby rail travelers were medically examined and their suspicious articles disinfected. Authorities recorded the local addresses of those travelers granted permission to enter towns in order to monitor their whereabouts.[64]

The Government of India continued to liberalize some of its plague policies, based on the Indian Plague Commission's recommendations.[65] It implemented the Commission's proposal that indiscriminate disinfection of travelers' persons and effects be abandoned, marking a reversal of its 1897 plague policy. Only those suffering from plague or with symptoms should be detained, and segregation of contacts and evacuation of infected areas were to be used cautiously and only when and where effective. House visitation and searches fell into '"general disuse because of the alarm and distrust which it [excited] among the people who, in their endeavours to evade it, sometimes spread the plague from one place to another."' The Commission concluded that the best plague policy was to interfere as little as possible with the sick, disinfect rooms frequently, examine the sick often, and encourage inoculation among other house inhabitants.⁶⁶

Plague control measures had been relaxed, in part due to public protest and in part because of their seeming ineffectiveness, but the Government of India remained extremely careful at the turn of the century about containing plague because of persistent fears of international repercussions. The protection of India's trade continued to motivate the Government of India's plague control policies, and pilgrimage continued to be a main target of those policies. While the all-India ban on pilgrimage had been lifted in 1898, the government continued to impose bans selectively. In 1900 the Government of India prohibited pilgrimage from the entirety of the Madras presidency even though plague was endemic only in the Salem district. Based on prior experience, the Government of India believed that more districts were likely to become infected by the onset of pilgrimage, but, more importantly, the 'highly infected condition' of the Bombay Presidency and Mysore State rendered it 'necessary to regard Southern India as an area from which the pilgrimage ought not, in deference to the feelings and wishes of Foreign Governments, to be permitted'. In response to its fears about plague spreading within India and to European and Turkish concerns of plague being communicated abroad, the Government of India issued a resolution under the Epidemic Diseases Act III, prohibiting pilgrimage from the Bombay Presidency, Calcutta, the Saran district of Bihar, and the states of Mysore and Hyderabad.⁶⁷ The Government of India noted that the 'alarm experienced by European Governments, including Turkey, regarding the danger of plague being imported to Europe is in no measure abated'. During the past pilgrimage season, Turkish authorities had again subjected Indian pilgrims to 'much vexatious restriction and consequent hardship' in Turkey's efforts to reduce the number of pilgrims. Pilgrimage from uninfected parts of India was still permitted, but the Government of India continued its surveillance of pilgrims leaving and re-entering India.⁶⁸

While British and Indian desires to maintain flows of trade had, at times, provided a powerful obstacle to international consensus on disease control and had limited the impact of earlier sanitary conventions, the metropole and the colonial state had come to appreciate by 1897 that observance of the international sanitary conventions could actually keep Britain's India trade running smoothly. The Government of India discovered that trade in Calcutta had not been significantly disturbed by the appearance of plague in the city in 1898, largely due to foreign countries and India both abiding by the terms of the 1897

Convention. The Government of India believed that adherence to the conventions by other states was predicated on the '"fulfillment in India of the obligations laid down by the Convention on infected countries. Should the belief spread [that] the requirements of the Convention are not being faithfully discharged in India, the injury to trade may be far greater than any that could be caused by the delay and inconvenience of disinfection on shore."'[69] Britain and India had found a way to balance their interests with international cooperation in disease control, for beneath the rhetoric of collaboration and collective progress ran strong currents of economic and political interest.[70]

## India's disease exceptionalism

While trying to reassure Europe that disease would not spread outside of India, the Government of India was also keen to negotiate more favorable terms for India in the international sanitary conventions. Its avowal of Indian exceptionalism in terms of disease ecologies and sanitary conditions allowed the Government of India to circumvent certain provisions in the conventions. Convention delegates from India argued that the international sanitary conventions had all been drafted from the perspective of protecting European health and trade, and had been 'framed and adopted by the European States to meet circumstances very different from those which prevail in this country'.[71] At the 1894 International Sanitary Conference, the British delegate, Thorne Thorne, tried to aid the Government of India's cause by proposing that the conference focus on the holy cities instead of on India as the primary sites of sanitary control. He argued that the eradication of cholera in British India was out of the question due to the size of India's territory and population (nearly 300 million); the fact that most Indians lived in thousands of isolated villages and not in cities; the difficulty of convincing people to change their hygienic practices; the number of Indians living in poverty; and the enormous cost of comprehensive public health measures. Thorne Thorne cleverly attempted to shift attention away from India by encouraging the committee to view cholera as permanently, and thus hopelessly, embedded in India's sanitary landscape. He proposed that pilgrimage regulation focus on the point of debarkation, the advantage being that sanitary surveillance would be limited to four cities – Jeddah, Mecca, Medina, and Yambo – with populations much smaller and less dense than those in India. A change in the locus of sanitary control would help alleviate the pressure on public health officials to reduce cholera in India amidst increased international scrutiny.[72]

The Government of India argued that while it did have public health obligations to other states, it could not be held to the same standards as Europe. Delegates representing India often declined to sign conventions, whose stipulations would have left India politically vulnerable, since the Government of India would have been hard pressed to follow the letter of the convention. For the most part, European delegates understood India's particular circumstances and made allowances for the conditions in which the Government of India had to operate. When the Government of India chose to decline ratification, it still implemented measures to abide by as many of the conventions' provisions as possible. The government wished to impress upon other states that it took seriously the task of containing disease, and in so doing, aimed to avoid cumbersome foreign restrictions on India's trade.[73]

India's exceptionalism at the International Sanitary Conferences was further facilitated by the peculiarities of sovereignty within the British Empire, which inadvertently created a loophole for the Government of India. Although India was a colony subject to Her Majesty's Government, it possessed its own administrative and political apparatus, which was subject to Whitehall, but not wholly dependent on it. In terms of public health diplomacy, the ties between Britain and India were less explicit and at times could be relatively weak. At the 1893 International Sanitary Conference Britain had signed the convention on behalf of the United Kingdom alone, expecting that India would follow suit, but India refused to oblige.[74] The British Under-Secretary of Foreign Affairs then sent a telegram to the Under-Secretary of State for India advising that British colonies accede to the convention, so that "'their abstention may not be urged by any foreign Government as a reason for declining to carry its stipulations into effect.'"[75] The 1893 convention would have required that the Government of India notify all signatory parties of the appearance of any concentration of cholera cases in India. India's particular cholera epidemiology meant that the disease could be epidemic in one area while a region only 100 miles away remained untouched. The notification of every outbreak would make India perpetually liable to quarantine by every European state, even if there were no cases in and around seaports. The Government of India declined to sign the convention, claiming that it would "'not only impose a needless burden on our administration, but would lead to results which, whilst securing no advantage to Europe, would almost certainly be productive of injury to this country'".[76] By exercising certain rights of state sovereignty, namely the right to ratify conventions, the Government of India was able to

adjust international expectations of its health policies, while reinforcing a certain degree of independence from Britain in terms of health
policy.

The Government of India could not indefinitely escape its colonial status, despite India's deputation of its own delegates to the International
Sanitary Conferences. The metropole was not always so accommodat-
ing. Just one year after the 1893 contest between the metropolis and the
colony, British delegates decided to accept the 1894 Paris International
Sanitary Convention on behalf of Britain *and* India. Britain approved the
convention, which had been drafted in response to the most fatal outbreak of cholera ever recorded on the *hajj*. Britain agreed to the compulsory inspection of pilgrims embarking from infected ports. Furthermore,
according to the agreement, India would have to accept the sanitary
reforms on pilgrim traffic to the Hedjaz imposed by the Constantinople
Sanitary Council, in spite of the Government of India's objection to
the Council's jurisdiction.[77] Surgeon-General J. M. Cuningham, the
delegate from India and former Sanitary Commissioner, told the conference that local circumstances in Indian ports were such that it would
be dangerous to detain pilgrims for five days.[78] It would be extremely
difficult to supply observation facilities at all ports for pilgrims entering the port area on different days. Furthermore, detention constituted
an infringement on pilgrims' liberty; he explained that the Muslim
population would be distraught at the idea. Gathering a large number of
people in the same locale, even for purposes of observation, would most
likely contribute to an increase in cholera cases before embarkation and
once on board the ship. He, thus, abstained from voting on the measure
to institute medical surveillance of pilgrims prior to embarkation.[79] The
Government of India found itself in the unenviable position of having
to weigh concerns for the religious sentiments of Indians against the
need to achieve consensus within the international community in order
to keep trade flowing smoothly.

British delegates listened but were not swayed by Cuningham's arguments. They told the Government of India that it was to the

> influence and pressure exercised by that Council that we look
> to procure the eventual fulfillment of the promises of reform in
> sanitary and other arrangements at the Kamaran quarantine station
> and in the Hedjaz. The due accomplishment of those reforms is of
> the utmost importance to the well-being of the pilgrims. And Her
> Majesty's Government did not deem it wise to disavow the jurisdic-
> tion of the Constantinople Sanitary Council in these matters.[80]

India's continued refusal to ratify the convention piqued British officials, who then forced the Government of India to incorporate the inspection of pilgrims into India's Pilgrim Ships' Act of 1895, overriding India's objections to the provisions of the 1894 convention. Indian port authorities and shipping companies would be required to medically examine pilgrims before embarkation and provide better living conditions on board. The act, as Britain must have hoped, helped restore international confidence in India and in the safety of Indian shipping.[81] To protect commercial interests, Britain made concessions and forced the Government of India to obey the spirit of the convention, if not its letter.[82]

Tension between the Government of India's capabilities and Britain's desire to protect shipping constrained India's ability to negotiate more favorable convention provisions. India and Britain were in precarious positions regarding international sanitary diplomacy. They needed to balance rejection of measures that would be logistically impracticable in India with compromise, which could ensure that Indian commerce proceeded unimpeded. Officials in India also had to proceed cautiously because they feared that any proposed amendments on technical matters could open the door for other proposals, which might disadvantage India and restrict freedom of navigation.[83] For this reason, Lord Curzon rejected the Government of India's proposal to conduct examination of passengers on board, which would be logistically easier, instead of on shore, as stipulated by the convention. Curzon reminded the Under-Secretary of State for India that allowances obtained at the Venice conference for Indian passengers and trade were better than had been expected, given the extreme reluctance of Mediterranean powers to support the British position. Concessions to British commerce regarding passage through the Suez Canal had already exceeded allowances granted at previous conferences.[84] Delegates for India, thus, had to employ the exceptionalism argument carefully and only when it would reap significant benefits.

Delegates' attempts to modify conventions in India's favor would continue throughout the life of the International Sanitary Conferences. By the early 1900s, the Government of India had proven its commitment to prevent the spread of epidemics from its territories, enabling it to obtain more diplomatic prerogative. At the 1903 International Sanitary Conference, it declined to subscribe to the cholera provisions of the convention because, given the logistical difficulties of reporting disease statistics for all areas of India, its terms would have rendered India a perpetually 'infected' country. The Government of India argued that there existed a 'marked distinction between the interior of India which comprises several different countries (and therefore is not analogous to a single European state) and the seaports which are in communication with Europe. As regards the interior we consider that it is not possible for us to accept any obligation to notify cases of cholera or to place the disease upon the same footing as plague'. With respect to seaports, the Government of India explained that the communication of weekly reports of the Boards of Health to foreign consuls and sanitary authorities in Europe already complied with the convention.[85] Furthermore, officials were stationed at Indian ports to detect infected persons and prevent their departure from the country.

## International health politics and domestic policy

International sanitary conventions had a significant influence on domestic policies in India, even if they did so in a sometimes counterintuitive way. The Government of India's arguments at the International Sanitary Conferences to obtain special allowances did not always square with its domestic practices. At the conferences, delegates for both Britain and India consistently condemned international quarantine, but the Government of India could be opportunistic domestically, ignoring principle in the face of potential trade loss. The Government of India imposed quarantine in June 1894 at ports in Bengal, Madras, Bombay, and Burma against ships arriving from plague-infected ports in Hong Kong and China. Merchants were anxious to ensure that the import trade did not place India at an increased risk of disease, even while they were keen to avoid quarantine against ships leaving India.[86] Two years later, when European and Middle Eastern ports began implementing quarantines against arrivals from Bombay, the Government of India, in cooperation with presidency governments, decided to impose quarantine against Bombay at the five major ports in India and at any subordinate ports in the Bombay Presidency that were in direct communication with Europe.[87] The Government of India chose to rely on self-imposed quarantine, however unpalatable, to allay overseas concerns about plague spreading from India. The Government of Bombay went one step further and utilized both land and sea quarantines during the plague outbreaks of 1896. The Bombay Plague Committee argued that land, including rail, quarantine, was even more necessary than sea quarantine, since symptoms had more time to manifest during the course of a sea voyage than during a rail one. These policies illustrate just how important European opinion and the protection of Europe were in structuring the Government of India's plague policies.[88]

Ironically, European authorities attacked the application of some of the very measures that had enabled India to prevent the spread of plague beyond its borders. European delegates to the International Sanitary Conferences criticized inconsistent plague policies and the lack of uniformity in quarantine regulations within India and throughout the British Empire. European states contended that England was an 'extreme quarantinist' whenever the British Isles were not concerned. The 'absurdity of the situation [reached] its climax in India', where presidencies often had different sets of quarantine rules and the Government of India showed inconsistency, at times temporarily reversing its anti-quarantine policy.[89]

British officials realized that these inconsistencies were inadvisable; it was illogical for British delegates to the sanitary conferences to ask for more liberal quarantine rules when so much variation existed within the empire, with some colonies enforcing measures more drastic than those of any Continental powers. Britain pressured India to align its quarantine practices with the rest of the empire in areas where they conflicted.[90] However, Indian trade came first and public health solidarity with the empire second. The Government of India and the provincial governments would do what was necessary to prevent future trade embargoes.

In order to implement these policies, India needed a domestic counterpart to the international sanitary conventions. To legalize any preventive measures necessary to contain plague, the Government of India passed the Epidemic Diseases Act, III in 1897. This act aimed to 'arm Government with power to prevent the carriage of epidemic diseases to foreign countries by infected passengers from India'. It empowered the 'Governor-General in Council when satisfied that India or any part thereof is visited by, or threatened with, an outbreak of any dangerous epidemic disease, to take special measures over and above those permissible under the ordinary law, with the object of preventing the outbreak of such disease and the spread of it'.[91] It also invested provincial governments with the power to prescribe measures to prevent the spread of epidemic diseases not only inside the province but also in other parts of India and to other countries.[92] The act passed with the intention that regulations for epidemics would be subject to general control of the Governor-General in Council, but regulations would generally be made by provincial governments with their greater knowledge and experience of local conditions and circumstances and with their greater abilities to gauge local opinion and enlist local sympathies. The Government of India utilized the act in four circumstances: prohibition of the Mecca pilgrimage; prohibition of emigration from India, especially from infected areas; suspension of the booking of rail tickets to certain localities with the goal of preventing large religious gatherings; and banning of the export of certain articles liable and likely to carry plague infection from the Bombay Presidency to other parts of India.[93]

The Epidemic Diseases Act granted the Government of India the authority to handle disease outbreaks in a manner that would meet with European approval. Even though the provisions of the 1897 convention had largely obviated the need for self-imposition of quarantine or bans on mass movements, the Government of India continued to resort to measures that went beyond the letter of the convention – an ironic behavior given its tendency to use India's disease exceptionalism to obtain greater leniency internationally.[94] When plague broke out again in 1898, the Government of India declared that Karachi was an infected port and therefore could not be open to pilgrim traffic, even though the 1897 convention had not prohibited embarkation at infected ports. The Government of India thought that no matter how careful authorities were in segregating and embarking pilgrims, '"Foreign Governments would certainly view with disfavour and alarm the departure of pilgrim ships from an infected port."' The government hesitated to issue another ban on pilgrimage to the Hedjaz, so, instead, it urged the Government of Bombay to utilize District Magistrates, Muslim religious leaders, and pilgrim brokers to dissuade pilgrims from going to the Hedjaz the following season.[95] In 1899, with plague still present in Calcutta, the Government of India issued orders for Aden, Rangoon, and Madras to impose the 1897 Venice regulations against Calcutta's ports. The Government of India was willing at times to accept temporary disruptions within the larger Indian empire for the greater good of Indian trade, pilgrimage, and India's international standing. What it was not willing to accept was another round of quarantines and embargoes by foreign governments.

The international sanitary conventions, in the hands of the Government of India, also became a useful tool with which to arbitrate internal disputes over domestic sanitary practices.[96] Both the Government of India and provincial governments cited conventions to justify disease control measures, and the Government of India relied on conventions to help mediate provincial and intra-imperial disagreements. When the Government of India requested that the Government of Madras stop placing passengers arriving from infected ports within the Indian Empire under observation for ten days (the fixed average incubation period for plague) in cases in which passengers had traveled on a healthy ship but without a medical man on board, it pointed out that the 1897 Venice Convention did not require this measure. The Government of Madras' Sanitary Commissioner, Walter Gaven King, supported his decision to place individuals under surveillance by also citing the Venice Convention, which permitted local authorities to use discretion in the treatment of crews and passengers. King felt that passengers sailing on ships without doctors should undergo observation because it would be impossible for the Port Health Officer to rely on laymen descriptions of disease symptoms. Furthermore, he did not see any reason to give preferential treatment to passengers arriving on sea versus those on rail, who were subject to observation if arriving from infected areas. Since surveillance aimed at the 'maximum protection of a population from importation of plague compatible with the least feasible disturbance of trade and passenger traffic', King advocated treating inland rail traffic and coasting traffic (meaning maritime traffic along Indian coasts or between parts of the Indian Empire) the same. The Government of India disagreed, arguing that the Venice Convention stated that passengers on healthy ships must not be detained for more than ten days from the date of departure from an infected port regardless of the presence of a medical officer. The Government of India concluded the matter by overruling the Government of Madras's order.[97]

The conventions proved just as useful when mediating disputes within the larger Indian Empire. Before the signing of the 1897 convention, the Government of Ceylon had subjected Bombay ships to a 15-day detention period to protect Ceylonese ports from plague in an effort to ensure the continued acceptance of its tea exports by foreign governments. The Bombay Chamber of Commerce appealed to the Government of India to reason with the Government of Ceylon to retract its excessive detention period. The Government of India duly asked Ceylon to bring its quarantine practices into line with convention regulations, which stipulated ten days' detention.[98] The Government of India emphasized the 'extreme importance of endeavouring to ensure that the trade of India shall not be subjected in British Colonies to regulations more severe than those which can be imposed by foreign Governments which ratify the Convention'. The Government of India also appealed to the British government for support but received no help. Britain confirmed that the convention would apply immediately to India and the Straits Settlements only; other British colonies and possessions would be encouraged but not obligated to adhere. (In this case, this loophole did not work in India's favor.) The Government of Ceylon argued that it needed to take such stringent measures to protect

Colombo, its only major port, because of the lack of knowledge about the nature and spread of plague. As 'trade with Europe was far more important to this Colony than trade with Western India, it was decided to sacrifice the latter in order to secure the former'. The policy proved prudent. Russia's prohibition, for instance, of imports from India was not extended to Ceylon. India was able to work out a compromise with Ceylon, which agreed to reconsider its detention period once the 1897 convention was ratified.[99]

These episodes reveal the extent to which burgeoning international health politics influenced Indian disease control policies. The Government of India applied international conventions to domestic and intra-imperial maritime traffic and wanted to ensure that conventions were followed to the letter within India. In reality, the government at times inconsistently applied the conventions, tending toward greater stringency, when it deemed it expedient. Because international sanitary conventions were critical to the regulation of disease control both within and without India, they became tools that could be employed by either the central or provincial governments in contestations over public health authority.

However, Britain's position of allowing colonies not to ratify the conventions undermined the Government of India's efforts, and by extension Britain's, to protect the India trade. With India under tight scrutiny by states conducting trade with it, Britain and its extended Indian Empire needed to present a united front. Dissension within the ranks was not to be tolerated, yet it repeatedly was. The arbitrary policies of different governments in the Indian Empire weakened India's position at the sanitary conferences. Once the 1897 convention had invalidated the practice of quarantine, the imposition of quarantine either within India or in other parts of the Indian Empire no longer made political sense. Rather, the Indian Empire as a whole should have rejected quarantine to protect imperial trade, but the Government of India faced a difficult challenge in coordinating governments across the Indian Empire. In May 1898, agents for the British India Steam Navigation Company and the Asiatic Steam Navigation Company complained to the Government of India that Rangoon port authorities, after the appearance of plague in Calcutta, had begun imposing a ten-day quarantine on all steamers from Calcutta regardless of the health of the ship and had refused to permit healthy ships to conduct business. Company agents complained that this practice contradicted the Venice convention and asked the Government of India to demand that the Government of Burma act in accordance with the international sanitary convention. Agents also complained that the Government of Madras had similarly imposed a ten-day quarantine against arrivals from Calcutta, practically shutting off steamer communication with Madras. Bird and Company asked the Government of India to urge the Government of Madras to comply with the 1897 Venice convention, especially since their steamers were carrying urgently needed coal for the South Indian Railway, Madras Railway, and Mysore Gold Mines. The Government of India ruled that trade from Calcutta to Burma was not to be considered coasting trade, and therefore, the special rules for coasting trade under the Venice convention did not apply. The Government of India reaffirmed its opposition to both land and sea quarantine, since they only injured trade and communication without conferring protection against the spread of epidemic disease.

> When, as in the present case, susceptibilities of other nations are not involved the Government of India are entirely averse from any departure from the general principle that quarantine should not be imposed. Acceptance of this principle is of the first importance for British trade in general and Indian trade in particular, and the Government of India could not consent to its infringement by one part of the Indian Empire against another. Such a course would probably be most prejudicial to general Indian trade interests, and might lead to a re-opening of questions settled by the Venice Convention.

The Government of India reprimanded the Madras and Burmese governments and ordered them to remove their ten-day quarantine against arrivals from Calcutta.[100] What is interesting about the Government of India's statement is that it left open the possibility of implementing quarantine should the 'susceptibilities of other nations' come into play.

The Government of India's fastidious adherence to the sanitary conventions, especially to the 1897 one, generally benefited India's trade and its standing in the international health community, but it could also lead to horrific consequences. When an outbreak of plague in Jeddah left indigent Indian pilgrims stranded in the Hedjaz, the Government of India declined to authorize a steamship to convey Indians back to Bombay, since the ship was not properly fitted for pilgrim traffic according to international regulations. Authorization of an inadequately equipped ship would have violated the Pilgrim Ships' Act and was out of the question given the British Government's recent ratification of the 1894 Paris Convention. Unwilling to contravene the Pilgrim Ships' Act and disinclined to pay for the repatriation of 600 indigent pilgrims, the Government of India and the British Government refused to accept responsibility for these pilgrims, allowing them to perish from 'exposure and want' rather than have Britain and India incur the negative repercussions of breaching international convention.[101] The Government of India made arrangements to convey those pilgrims who could afford the return trip to India but abandoned indigent pilgrims in Jeddah.[102] Although the Government of India willingly protested international sanitary regulations that would have infringed on Indian Muslims' 'liberty', this particular group of Indian Muslims was not afforded protection. The nature of empire meant that the political and economic priorities of the colonial state would repeatedly take precedence over its subjects' health and safety.

## Conclusion

The epidemiological risks that plague and cholera posed to Europe and the economic and political risks that not containing the epidemics posed to India fundamentally shaped the Government of India's policy responses to the plague and cholera epidemics of the 1890s. During plague and cholera outbreaks, the Government of India instituted the requisite measures to reassure Europe that disease would not spread beyond India's border. In doing so, India ensured the continued flow of its trade and left intact political and economic relationships among Britain, India, and Europe and within the British Empire.

The role of British India in propagating epidemic disease around the world had raised questions at the International Sanitary Conferences about the relationship between state jurisdiction and self-interest and a state's public health responsibility to a larger international community. That responsibility was predicated on two assumptions: the interconnectedness of vast and far-flung regions of the world through trade and travel facilitated the spread of epidemic disease; and international cooperation was necessary to reduce the level of risk to the health of populations and to the functioning of the global economy. The use of trade embargoes and the political and commercial repercussions of violating conventions became a useful tactic for European states to force Britain and the Government of India to recognize that they, too, had to accept limitations on their sovereignty in the interests of the international community and for their economic self-preservation.[103] The economic and political risk for India if it did not abide by conventions was too high. So although the Government of India at times made reservations to certain convention provisions, the international sanitary conventions had circumscribed the space within which the Government of India could maneuver.[104]

There was another type of risk that the Government of India had to consider – risk to its political relationships with its subjects. The central and provincial governments needed to maintain political and social order and avoid fueling discontent within large sectors of both the Hindu and Muslim populations. They also needed the collaboration and cooperation of Muslims as a counter to a growing Hindu political class. As a result, the Government of India had to weigh the international risks against the domestic ones in choosing how to respond to plague and cholera outbreaks. In the end, the international ones proved more important.

The Government of India's greater involvement in disease control in India and internationally had resulted from the coalescence of different sanitary regimes in Europe around the prevention of cholera and plague. The international sanitary conventions, regardless of whether India signed them, had the authority of international consensus, and thus pushed India to take more decisive steps to contain the spread of cholera and plague. However, by virtue of their scope and authority, the conventions could only address diseases that threatened to cross international borders. Endemic diseases, so long as they did not morph into epidemic outbreaks, lay outside the conventions' purview and Europe's interest. While Europeans understood the connection between their health security and the presence of epidemic disease in distant lands, they had not yet made endemic disease in their colonies their concern, except where and when it interfered with colonial expansion and development.

In keeping with a Westphalian conception of international public health, Europe did not interfere in India's endemic disease landscape. This had the unintended consequence of allowing the Government of India to focus on containing epidemics in moments of crisis to the detriment of sustained endemic disease control. The Government of India's emphasis on epidemic disease meant that fewer financial, administrative, and medical resources were available for the prevention of endemic plague, cholera, and malaria. These three permanently embedded diseases exacted a far greater annual toll than did periodic epidemics in India.[105] This marked distinction between epidemic and endemic disease would prove of utmost significance in the Government of India's approach toward plague and cholera throughout the first half of the twentieth century.

# 2
# Plague and Cholera – The Epidemic versus the Endemic

Between 1892 and 1940, over ten million people died of plague in India and over 15 million died of cholera. Plague and cholera instilled fear in their victims and in anyone who might come into contact with them or their effects. Indian port cities, teeming with undernourished populations housed in overcrowded and insanitary dwellings and without access to pure water, could not have been more welcoming of disease. Bombay, in particular, was ripe for plague infestation. It hosted an abundant and susceptible population of black rats, whose traveling companions, *Xenopsyllia Cheopis*, happened to be one of the best flea vectors for plague transmission. Modern transport, the grain trade, and the close proximity of rodents, fleas and humans in Indian homes disseminated plague and its vectors through Indian towns and villages. The impoverished were especially susceptible, since they could not afford to build the sanitary dwellings that would shelter them from rats and their plague-carrying rat fleas. Cholera, too, found its home in particular groups: the poor, who were especially vulnerable during famine or scarcity; migrants searching for work or food; and pilgrims.[1] Plague and cholera fed on the economic deprivation, dense and unhygienic living conditions, and chronic malnutrition that afflicted the poor.[2] Those living in rural poverty, unlike some urban dwellers, lacked access to the few piped water systems being constructed in urban areas.

Unfortunately for the millions of Indians afflicted with plague and cholera, the Government of India and Europe were more preoccupied with the spread of epidemic disease beyond India than with endemic disease that stayed within India's borders. The ratified international sanitary conventions had not required states to control disease or improve sanitary conditions domestically.[3] High levels of mortality had no direct effect on the continuance of international trade, even though they undoubtedly hurt India's productivity and devastated families and communities throughout India. Public health officials in India well understood that epidemic cholera and plague had markedly different economic and political ramifications for the Government of India than did endemic forms of the diseases. The forms of these diseases – endemic as opposed to epidemic – acted as critical variables in the determination of policies in India. Policies reflected the central state's and European perceptions of differences in risk posed by the endemic versus the epidemic as well as the interests of the parties who had a stake in the Government of India's disease policies.

## Epidemic disease

### Economic imperatives of empire

During the plague and cholera epidemics of the 1890s, European states were not alone in advocating decisive disease control measures in India. During the 1896–7 plague outbreak, Europe and the 'official mind' in both London and Calcutta found common cause with investors, traders, and businessmen in India and Britain. The horizontal center–periphery relationship collided with the vertical merchant, 'gentlemanly capitalist', governor, viceroy, ambassador relationships to spur the Government of India into action.[4] As merchants began feeling the economic fallout of the outbreak of plague in Bombay in 1896, mercantile associations, such as the Indian Tea Association and local Chambers of Commerce, pressured the Government of India to institute drastic measures to limit the commercial damage resulting from plague-induced trade embargoes. When plague struck Bombay and Karachi, Russia had prohibited the import of Indian tea into Batoum from Calcutta and followed this prohibition by banning the importation of all Indian merchandise from all Indian ports into Russia. The Bengal Chamber of Commerce believed that the Russian Government's steps were prompted by the belief that the Government of India had not taken sufficient action to limit the spread of plague. The chamber warned that if

> other nations of Europe were to follow the example set them last year by Russia, and this year prohibit the import of all Indian produce from all ports, the results to the export trade of Bengal would be disastrous . . . . [it would be] hardly necessary for the Committee to touch upon the disaster which would follow, on Calcutta being closed as an export port, to the agricultural, planting, and manufacturing

industries of Bengal and Assam, and finally to the Government itself, owing, for example, to the inability of the growers of produce to pay land revenue if there were no market for their produce . . . . any further steps which can be taken to materially reduce the probability of the plague reaching this Presidency will meet with the entire approval of the mercantile community.[5]

India's trade associations used their powerful voice to lobby the Government of India to implement stringent disease control measures to ensure continued commerce.

The Government of India's willingness to listen to trade lobbies stemmed from its desire to protect trade flows and reflected the growing role that India had in London's management of the international economy.[6] The Government of India wished to be seen as responsible when it came to ensuring the health and economic safety of international trade, but when and where international trade was not involved, the Government of India was more flexible. When petitioned by mercantile associations or mining and railroad companies, the Government of India made exceptions to its policies in order to protect domestic labor supplies and intra-imperial labor flows that supported Indian and imperial economies. Merchants and companies wanted to mitigate any economic damage resulting from the policies the Government of India had had to enact to satisfy delegates at the sanitary conferences, and the government was receptive to their suggestions.

The Government of India viewed laborers (both migrants and emigrants) as a special class within the population and applied different rules to them in terms of medical examinations, detention, and freedom of movement. In the 1890s and early 1900s, increased labor migration from rural to urban areas and to Assam tea plantations had facilitated the spread of epidemic and endemic diseases.[7] Although the Government of India generally condemned detention of ordinary passengers, especially in camps, because it caused hardship, provided opportunities for extortion, and, furthermore, failed to check the spread of plague, it supported the detention of laborers. For those traveling to work in the tea gardens, detention was not only permitted but encouraged in order to protect the health of the labor supply and to prevent migrating laborers from spreading plague further.[8] The last thing the Indian Tea Association needed was for its labor-intensive industry to be brought to a halt because of plague-infected laborers.

Plague placed the labor supply and hence the commercial and economic viability of India at risk. Both the Government of India and trade associations wanted to prevent disease spread among laborers within India. But Indian labor was also critical to the economy of other parts of the British Empire around the Indian Ocean. Between 1895 and 1914, roughly 39,000 indentured laborers and 10,000 voluntary migrants traveled from India to work in Kenya and Uganda. The number of Indians who journeyed to Lower Burma, primarily to work on rubber plantations and in tin mines, increased from 297,000 in 1901 to 583,000 in 1931. Annual migration to Malaya grew from 15,000 in the 1890s to 90,000 in the 1920s.[9] Labor flows tied British possessions in India and in Africa together, and, if the imperial economy in the Indian Ocean was to continue functioning, India could not ban all labor emigration in an effort to control plague.

The Government of India's plague regulations had to take into account the needs of other parts of the empire, which necessitated finding legal loopholes in the Government of India's legislation. The government had ordered that no British or Portuguese Indian native resident in or passing through territories in the Bombay Presidency since 1 January 1897 was permitted to embark at any British Indian port for a port outside of British India.[10] Therefore, overseas enterprises had to recruit laborers outside of the Bombay Presidency.[11] While the Government of India opposed ordinary emigration from Karachi while the city was deemed infected, it requested on behalf of the British Government, which was building the Uganda Railway, if the Government of Bombay would permit servants, artisans, and laborers engaged to work on the Uganda Railway or under the Uganda Protectorate to proceed to Africa from Karachi if they had first been segregated for ten days. The provision of Indian labor was critical to the railway's progress, since construction was primarily undertaken by Indians. The Government of Bombay replied that the Commissioner for Sind could arrange for emigrants to Mombassa to be segregated. The Government of India then permitted these classes of people to emigrate after satisfactory medical inspection.[12] Similarly, when the Seychelles Government asked the Government of Bombay for 200 masons, miners, and laborers for road construction, the Government of India responded that if Seychelles was willing to take the risk, then the rules could be relaxed as long as men were isolated properly for ten days and their clothes disinfected before embarking.[13] These exceptions to general policy, representing two among many, served the economic interests of the British Empire. There was one caveat: the Government of India would only grant these allowances when and where they did not place Europe and Turkey at risk for disease.

## Pilgrim politics

Laborers were not the only special class of people where infectious disease policy was concerned. The International Sanitary Conferences had identified other, special classes of people and legitimized their regulation. The 1897 Venice Convention allowed governments to take measures targeting gypsies, vagabonds, emigrants, and pilgrims. Participating states viewed large masses of mobile people as disseminators of disease and, therefore, subject to regulations not approved for the general population. Pilgrims, both those traveling to religious fairs within India and those voyaging to the Hedjaz, constituted a dangerous category of 'border crossers'.[14] The control of pilgrimage and of pilgrims, thus, became central to both international and Indian public health policy.

Delegates at the International Sanitary Conferences had long viewed pilgrims as special threats to public health. In 1891, W. J. Simpson, Health Officer of Calcutta, argued that Europe need not fear the importation of cholera from trading ships traveling from India to Europe through the Suez. Cholera was not prevalent among the Indian workers and Europeans on commercial ships. Rather, the danger came from pilgrimage, which transported classes of people, often poor and living in insanitary conditions. His argument was flawed in that trading ships carried pilgrims as well. He contended, however, that Mecca and Jeddah were the most dangerous centers of cholera as far as importation of cholera by sea into Europe was concerned. Simpson argued that if European states only realized this, it would greatly benefit trade between Europe and the East and reduce unnecessary detentions and quarantine restrictions.[15] Simpson tried to shift the focus of sanitary control from trading ships to pilgrimage, while simultaneously advocating that Europe impose requirements for sanitary surveillance of pilgrims who had already arrived in the Holy Lands, instead of on those en route. Turkish authorities were particularly hostile to Indian pilgrims because they, like Europe, feared the spread of cholera and plague into their territories. Unfortunately for the Government of India, Simpson's viewpoint did not prevail at the International Sanitary Conferences; pilgrims coming from India, not those already in the Hedjaz, would become the major focus of cholera prevention and of early plague control.[16] As a result, their sanitary regulation became an integral part of the Government of India's policies to safeguard India's trade.

In negotiating suspensions of pilgrimage, the Government of India realized it needed to address primarily the health concerns of European states and the economic ones of Indian mercantile communities. At the same time, it understood that it could not neglect the practices and opinions of the thousands of Indians that went on pilgrimage. Although pilgrims as a group were not critical to India's economy in the same way that laborers were, Indian Muslims were an important political constituency that the Government of India did not wish to alienate. They acted as a political counterweight to the growing influence of the Indian National Congress, formed in 1885, and to an assertive Hindu political class. And after the passage of the 1909 Morley-Minto Act, which created separate Muslim and Hindu electorates, Muslims became a distinct, political category and, hence, an even more important political constituency.[17] Although pilgrims did not have a powerful association to speak directly on their behalf, as did the mercantile sector, they were an important indirect influence on the Government of India's policies because of the government's politics of collaboration. The government, however, could not make the same exceptions for pilgrims as it had made for labor while a ban on pilgrimage was still in effect. The best the Government of India could do was to cushion the impact and divert blame away from itself.

In the days leading up to the 1897 Venice conference, the Secretary of State for India and the Government of India found themselves in an increasingly difficult position. They anticipated that pilgrimage restrictions would be proposed at the conference, but they had no strategy for responding. There was no consensus among provincial governments or mercantile organizations on how to restrict pilgrimage to the Hedjaz. The Secretary of State for India thought it advisable to stop pilgrimage from India, but the Government of India objected to a complete suspension of pilgrimage. On February 13, three days before the start of the conference, the Government of India polled provincial governments for their views on how Muslims would perceive a ban. The Government of Bengal agreed with the suspension of pilgrimage from Calcutta for the remainder of the year because of the danger in diverting large numbers of pilgrims from the port in Bombay to Calcutta; it thought that its Muslim population would understand. The Government of the North-West Provinces, Oudh strongly opposed prohibition of pilgrimage and felt its Muslim population was less inclined to accept a ban. The Lieutenant-Governor of Punjab thought, in agreement with the Government of India, that it would be better for India to stand firm against a ban and 'wait and let the measure be forced on us by the Conference. Possibly it could be arranged that the [Ottoman] Porte should take initiative in urging it', which would divert any animosity away from the Government of India. The Chief Commissioner of the

Central Provinces urged the Government of India to continue to protest suspension in order to avoid setting a dangerous precedent for future disease outbreaks. The Government of India, after weighing the views of provincial governments against current and potential restrictions on Indian trade, decided on 16 February, while the Venice conference was in session, to prohibit the departure of pilgrims resident in the Bombay presidency and Sind and to close the Bombay port to pilgrims.[18]

Mercantile communities, who were attentively following these discussions, were not satisfied with this restriction. Trade associations believed that a ban on pilgrimage from all of India was a necessary prerequisite to the lifting of international trade restrictions and so lobbied the government to implement more stringent measures. They objected to banning pilgrimage from only the Bombay presidency, as this would divert pilgrims to other ports. Mercantile organizations, along with the Government of Madras and the Government of Bengal, began pressuring the Government of India to close all ports in India to pilgrimage. The Indian Jute Manufactures' Association and the Indian Tea Association protested the diversion of pilgrims to Calcutta, which they argued would increase the risk of infection, and hence the likelihood of quarantine – a risk the city could ill afford given the large proportion of India's trade passing through Calcutta. The Bengal Chamber of Commerce worried that large groups of workers traveling to and from the jute and cotton mills, the coal mines, and the tea gardens of Bengal and Assam would become susceptible to infection if brought into contact with 'gangs of pilgrims' traveling from western India. Infection of migrant labor could then lead to a swift spread of plague throughout the provinces.[19] The Lieutenant-Governor of Bengal made it clear that the question of an all-India pilgrimage ban turned on the relative political importance of Calcutta mercantile opinion versus Muslim religious sentiment. Only the Government of India could make that judgment, and trade associations had more of the Government of India's ear than did pilgrims.[20] Still, the Government of India was 'averse from passing orders which will have the effect of stopping pilgrimage and be regarded as interference with the religion of Muhammadans'.[21] On the other hand, the Bombay and Karachi ports had already been quarantined by many foreign governments, and the Government of India could not risk Calcutta being quarantined as well, since it was practically the only remaining outlet for northern Indian produce and goods.

In the end, the Government of India was spared the burden of deciding what to do; the final decision came from Whitehall. The Secretary of State for India telegraphed the Government of India that 'in consequence of strong opinions regarding possible communication of plague to Europe of all European Governments whom nothing short of prohibition will satisfy, Her Majesty's Government have decided that pilgrimage should be altogether suspended for the present season'. Viceroy Elgin realized that he had no way out, but he wanted to be as strategic as possible. He asked if 'all European Governments' included Turkey. If so, it would strengthen the Government of India's position that this was a measure being forced on it from two sides – Her Majesty's Government and the Ottoman Porte. The Secretary of State for India responded to the Viceroy that, 'Turkey insists as strongly as other nations on precautions against plague, but has not specifically pressed for stoppage of pilgrimage'. The Government of India carefully worded its notification to communicate to its subjects that its hands were tied. On 20 February 1897, the Viceroy of India, under orders from the British government, issued the following notification: 'The question of the suspension of the pilgrimage to the Hedjaz having been under the consideration of the Government of India and Her Majesty's Government, Her Majesty's Government have now come to the conclusion that, in consequence of the strong opinions of all European Governments, including Turkey, regarding the danger of plague being communicated to Europe, it is impossible to meet their demands by any measure short of the suspension of the pilgrimage for the time being'.[22] In November 1898, the Government of India would once again prohibit pilgrimage to the Hedjaz, but this time only from the Bombay and Madras Presidencies and from Hyderabad State.

The Government of India was well aware that its actions would upset Muslim communities, who might resent a curtailment of their ability to perform religious duties. It therefore sought the collaboration of elites to help persuade these communities of the necessity that the pilgrimage should be carefully restricted, in order to prevent plague being communicated to Europe and spreading from infected to non-infected areas of India.[23] The Government of India urged the Government of Bombay to utilize district magistrates, Muslim religious leaders, and pilgrim brokers to dissuade pilgrims from going to the Hedjaz.[24] In April 1898, the government even went so far as to obtain sanction from the Ottoman Caliph for the suspension of pilgrimage to the Hedjaz from European colonies.[25] The Sultan agreed to sanction measures to suspend pilgrimage if the measures were proposed by all European governments represented on the Constantinople Sanitary Council. The Caliph, however, later rescinded his sanction of the ban on pilgrimage, stating that it opposed the precepts of the Koran.[26]

Turkey's refusal to support the suspension or to ratify conventions placed the Government of India in a delicate position in relation to its Muslim subjects, especially given the Government of India's courting of India's Muslim elite as a counter to Hindu political leaders. From the 1890s through the fallout of World War I, Turkey had considerable influence on India's pilgrimage policies because pilgrimage sites were located within the Ottoman Empire and because the Government of India wanted to protect its Persian Gulf trade. The Government of India found itself in a difficult situation because both Shia and Sunni Indians wanted to go on the *hajj*, but Turkey was not equally welcoming of both groups. In 1901, the Constantinople Board of Health finally withdrew its prohibition to the entry of Indian Shia pilgrims into Turkey provided that ships did not carry more than five Shias per 100 tons of vessel register.[27] The Government of India objected to Turkey's sectarian classi-fication of pilgrims and disagreed with the limitation placed on Shias, but the Government of India did not think it advisable to protest this sectarian discrimination. Pilgrim traffic with Basra, the site of Shia holy sites, represented only a small portion of trade in the Persian Gulf, but if the Government of India protested the discrimination against Shias, Turkey might restrict entry of all Indian pilgrims. This in turn would seriously hamper passenger traffic, and thus trade, with the Gulf. The Government of India thought it better to avoid a general prohibition and protect the Persian Gulf trade rather than incur further hostile measures.[28] Furthermore, since Turkey had not ratified the 1897 con-vention, the Government of India could not protest restrictions on Indian pilgrims and quarantine of ships sailing from India.[29]

After the turn of the century, several factors led to an easing of plague control measures relating to pilgrims (although Turkey would continue to impose stringent restrictions). Advances in understandings of disease changed the public health interventions available to health authorities. Greater scientific consensus on plague etiology and its method of spread aided international agreement on what measures to take and to require of other nations.[30] Once scientists had determined the incubation period for plague and had discovered and elucidated the rat flea vector and rat host mode of infection, the Government of India could focus less on policing people and more on eliminating rats.

As the medical profession gained more knowledge about plague, modifications were made to the international sanitary conventions. With Europe's approval formally codified, the Government of India and provincial governments were able to reduce the period of detention and observation of pilgrims before their embarkation for the Hedjaz.[31] The

Bombay segregation camp for pilgrims was officially abolished in 1907, although medical inspection of clothing and bedding continued. After the Plague Investigation Commission's report of 1908 concluded that plague was transmitted by infected rats and their fleas, the Government of India decided that the existing system of medical inspection of passengers traveling on ships could no longer be scientifically justified.[32] Advances in etiological knowledge had shown officials that pilgrim movements had little effect on plague spread, even if they were still problematic for the diffusion of cholera.[33]

The Government of India also modified its plague regulations for reasons of domestic political expediency. The Government of India felt that it had acquiesced long enough in appeasing 'less enlightened' European powers. The ban presented a great administrative burden, elicited much protest from Indians, and was not actually required by the terms of the Venice convention. After a five-year ban prohibiting residents of the Bombay Presidency (excluding Sind) from going on pilgrimage to the Hedjaz, the Government of India decided to lift the ban in 1902 in response to the Government of Bombay's protests that this policy interfered greatly with Muslim sentiments. Furthermore, the Turkish Consul General had been telling Muslims in Bombay that the Ottoman Porte had had nothing to do with the restrictions placed on Indian pilgrims, thereby shifting responsibility for the suspension to the Government of India – a myth the Government of India wanted to dispel.[34] Following much protest from Muslims in Bombay, the Government of India, using the discretion authorized by the Paris Convention, abolished segregation in Bombay, choosing to rely on deratisation, medical inspection, and disinfection of pilgrims and their baggage.[35]

With all suspensions of pilgrimage removed, the touchy question of compulsory inoculations for pilgrims against both plague and cholera returned to the fore. The anti-plague inoculation had been widely regarded by officials in India to be efficacious, even if the duration of immunity was not permanent.[36] The Indian Plague Commission had reported that the anti-plague inoculation was shown to ward off plague and to decrease the fatality of an attack.[37] The popularity of anti-plague inoculations, especially in urban areas, continued to grow. But since scientific knowledge had shown rats and not people to be the diffusers of plague, anti-plague inoculation became less critical for pilgrims and for the Government of India.

Anti-cholera inoculation was a different matter, since people were directly involved in cholera's transmission.[38] The results obtained demonstrated a marked reduction in morbidity and mortality among the inoculated as compared to the uninoculated. The inoculation appeared to provide protection even after an epidemic had erupted. Unlike plague, cholera did not require a vector or animal host, making humans the focal point of disease control. Rather than simplifying, this complicated disease precautions. In 1916, the Secretary of State for India asked the Government of India to consider introducing a system of voluntary anti-cholera inoculations for pilgrims. The Government of Bombay advised the Government of India not to promote voluntary inoculation on several grounds: introducing another inoculation could jeopardize voluntary smallpox vaccination, which, while not generally appreciated as to its benefits, was at least tolerated, and active opposition had practically ceased; it might be difficult for pilgrims to recover from two inoculations before embarking on the trip; only a very small number of cholera cases were imported by returning pilgrims; and the duration of immunity from the cholera inoculation was short-lived.[39] Following the war, the Government of India also became more cautious about instituting restrictions on pilgrimage to the Hedjaz with increasing participation of Muslims in nationalist movements and with widespread Indian involvement in the Khilafat movement.[40]

Mandatory inoculation of pilgrims against cholera continued to be debated throughout the 1930s. The Government of India consulted the Haj Pilgrimage Committee of the OIHP and the Indian Haj Inquiry Committee, which both recommended that the Government of India provide compulsory vaccination against cholera at no charge for those pilgrims intending to go to the Hedjaz from India. The Government of India accepted their recommendations, at least in theory.[41]

In practice, the Government of India desisted from making inoculation legally compulsory but did employ other means for ensuring pilgrims' cooperation.[42] The 1926 international sanitary convention had exempted pilgrims who landed at the Kamaran quarantine station from undergoing unpleasant sanitary and disinfection procedures if all pilgrims on their ship received smallpox and cholera immunizations. (Based on the data available at the time, many in the international health community believed the cholera immunization to be efficacious.)[43] Smallpox vaccination was already customary in Calcutta, Karachi, and Bombay ports, and officials thought pilgrims likely to accept anti-cholera inoculations if it meant they could avoid unpleasant sanitary procedures at Kamaran. During the 1930 winter–spring Haj season, health officials made arrangements at Karachi for voluntary anti-cholera inoculations. In reality, 'voluntary' translated into the Government of Bombay asking shipping companies to embark only inoculated pilgrims and the Government of India agreeing to cover anti-cholera inoculation expenses.[44] For the 1931 Haj season, the Government of India requested that the Government of Bombay work with the Bombay Haj Committee to have shipping companies 'cooperate by declining to accept uninoculated pilgrims as passengers'. Pilgrims received a ticket only after being vaccinated, while permission to embark was granted only after authorities had performed a same-day medical inspection and disinfection of the pilgrim.[45] The Government of India thought it important to have the majority of pilgrims inoculated against cholera before they left India.[46]

The Government of India also concerned itself with preventing cholera outbreaks among pilgrims traveling to religious sites and festivals *within* India. Because of crowded and unhygienic conditions and nature of religious rituals, the all-India distribution of festivals and the periodicity of Hindu pilgrimages were major factors in cholera's epidemiology in India. Since mass movements created opportunities for disease diffusion, the sanitary control of domestic pilgrimages and religious fairs was necessary to prevent the eruption and spread of epidemic cholera within and beyond India. Religious fairs and pilgrimages within India operated in connection with local markets, creating a large network through which flowed large numbers of people and goods.[47] The annual numbers of domestic pilgrims far outstripped the numbers going on the *hajj*, making cholera control more challenging, but also more critical. Over 120 religious fairs in British India each had an annual attendance of over 50,000 pilgrims. Religious practices at holy sites were particularly problematic given that cholera is a water-borne disease. Self-immersion in sacred rivers or tanks and drinking unsafe water from the Ganges as part of purification rituals created ideal conditions for the rapid spread of cholera at pilgrim sites. Cholera was further disseminated by transporting contaminated Ganges water home to relatives.[48]

Health officials, all too aware of the connections between pilgrimage and cholera transmission, made vigorous efforts to prevent a cholera outbreak. Provincial health departments continually monitored their jurisdictions for the first signs of an epidemic and established permanent anti-cholera organizations to ensure the implementation of necessary precautions at religious festivals.[49] Religious gatherings became enclaves of targeted sanitary control, especially after advances in understanding the etiology of plague had rendered cholera the biggest risk during domestic pilgrim movements.[50] The Government of India and provincial governments showed remarkable and rare public health cooperation during the 1927 Kumbh Mela, a large Hindu religious gathering held every 12 years at Hardwar and usually attracting between one and two million pilgrims. The Government of India and the Government of Punjab carefully oversaw sanitary conditions in order to avoid a repeat of the cholera pandemic that had spread from Punjab across Afghanistan into Russia and on to Europe in 1892. The Government of India inspected the latrine, first-aid, and drinking-water provisions at the fair site, while the Government of Punjab constructed special camps, roads, and bridges to allow orderly processions, augmented the number of beds at infectious disease hospitals, established traveling dispensaries and temporary hospitals, and controlled the arrival and departure of pilgrims at rail stations.[51]

Interestingly, while the Government of India strongly encouraged and at times coerced pilgrims into getting anti-cholera inoculations before traveling to the Hedjaz, it adopted a different stance on cholera prevention and treatment at religious festivals within India, choosing to focus on sanitation over the promotion of inoculation. The Sanitary Commissioner with the Government of India worked with provincial pilgrim committees, formed in response to the Government of India's requests, to design schemes for sanitary improvements to prevent pilgrimage centers from becoming foci of epidemic disease. The Government of India also allocated funds to local governments to improve the sanitary conditions along pilgrim routes within India.[52] In 1929 and 1930 the Government of India and provincial governments again considered the legal, cultural, and financial implications of introducing compulsory cholera inoculation for pilgrims going to festivals and religious sites within India. Domestic religious gatherings and pilgrimages presented a problem not just for the immediate locality of the site but also for towns and villages en route to and from the destination, where pilgrims could pick up and spread cholera if passing through infected areas. Provincial governments, however, unanimously agreed that compulsory mass inoculation was impracticable in the current political climate; it would very likely encounter grave resistance and could threaten the peace in an already politically restive society fueled by Gandhi's civil disobedience movement.[53] They did agree that voluntary inoculation should be encouraged among pilgrims.

As public opinion of inoculation became more favorable, the Government of India was able to align its approach to cholera prevention during domestic religious fairs and pilgrimages with its strategy of heavily promoting inoculation for pilgrims traveling to the Hedjaz. Public support of anti-cholera inoculation had grown rapidly, albeit unexpectedly, since 1930, permitting compulsory inoculation to be reconsidered. In 1928, 3,427,846 people had received anti-cholera inoculations in British India. By 1938, aided by increased health expenditures from the provinces, the number of inoculations administered had jumped to 10,779,332.[54] Although the Government of India still refrained from implementing a policy of mandatory inoculation, provincial governments began instituting compulsory inoculation for pilgrims traveling to religious fairs within India.[55] An outbreak of cholera following the Hardwar festival in April 1938 refocused official attention on the need to take all known precautions against the disease at religious festivals. The outbreak led the Central Advisory Board to discuss whether preventive inoculation could be enforced for pilgrims traveling to fairs within India; however, compulsory inoculation remained at the discretion of provincial governments.[56]

## Research as policy

The subject of compulsory inoculation illustrates how medical research and technology shaped the Government of India's policies to prevent and contain epidemic plague and cholera by enlarging the spectrum of public health intervention options available to government officials. Medical research also became an integral part of the Government of India's plague and cholera policies for a very different reason. The advancement of medical science was a particular type of colonial project, one that fulfilled an important legitimizing role for both Britain and the Government of India within the international public health arena. By 1900 tropical medicine had become vital to legitimizing empire and to realizing empire's economic potential and profits. More effective treatments for tropical diseases enabled laborers to be more productive and empires to expand development into otherwise difficult disease environments.[57] Advances in disease treatment allowed colonial governments to offer an illusory promise of better health for its subjects, while tropical diseases research demonstrated to the world the benefits of empire to medical advancement.[58] It also served the Government of India's political agenda to make India a renowned authority in the international public health world.[59] Medical research and its formal embodiment in institutes and laboratories lent international prestige to Anglo-Indian as well as British scientific endeavors and expertise. As early as the 1890s, a sense of international competition helped fuel disease investigations in India. Officers and researchers in the Indian Medical Service lamented that England was '"lagging far behind in the matter of tropical diseases where she should lead"'.[60] India should not have to go to 'France to learn about malaria, to Norway about leprosy, to Germany for cholera, and to Japan for plague'.[61]

Epidemic outbreaks of plague were the impetus behind the establishment of India's first research institution.[62] The 1898 Indian Plague Commission appointed to investigate the epidemic and the anti-plague inoculation concluded that the inoculation needed further trial and more research. It recommended that the government establish research laboratories for this purpose.[63] The outbreak of plague in 1896 and the Indian Plague Commission's subsequent recommendations highlighted the need for organized medical research and led to the establishment of the Plague Research Laboratory at Parel near Bombay in 1899, which undertook the production of anti-plague inoculations under the direction of Waldemar Haffkine.[64]

The Government of India's policy of relying on other countries to advance medical research was becoming less viable as India's disease outbreaks received greater international attention. The founding of research institutions and the increasing number of etiological studies in India assured Europe that India would address its infectious diseases and helped cement India's expanding role in international public health circles. Oddly, the rhetoric of the civilizing mission was absent from discussions about establishing research institutions. Instead, the primary motivation seemed to be to win the race for scientific discovery among empires.

Secondarily, health officials also recognized the practicality of establishing research institutions in India. Since tropical diseases and it victims were to be found in India, not in Europe, it would avoid the extra expense and time of sending Indian medical experts to Europe for training in bacteriology and parasitology.[65] The government needed scientific opinions as to how disease was communicated, the causes of outbreaks in different areas of India, the value of prophylactic and therapeutic sera in treating plague and cholera, and the conditions affecting the spread and decline of epidemics. Explanations for these questions remained obscure at the turn of the century, but as research became more entrenched in the Government of India's health policies, answers slowly emerged.[66]

The institutionalization of medical research got off to a shaky start. In what would become a recurrent theme in infectious disease policy, the Government of India and provincial governments found themselves at odds. Provincial governments did not unanimously support research endeavors just as they had not unanimously supported the initial ban on pilgrimage. The Central Committee of the Pasteur Institute of India had hoped that the Government of Bengal would consider seriously the establishment of an institute in Bengal with a broad research scope that would encompass diseases such as cholera, kala-azar, beri-beri, and rabies, diseases that frequently affected plantation laborers. It urged the Government of Bengal to provide administrative and financial assistance to an institute, which aimed to prevent and cure disease and suffering throughout the 'Indian Empire'. The Lieutenant-Governor for the Government of Bengal responded that in a 'poor country like India, the Government should, as a general rule, refrain from active interference in matters which lie on the borderland between science and speculation' and should avoid large expenditures on research which Europe in due time would answer and reveal.[67] This opposition to funding research was difficult for the committee to understand given that the proposed research would benefit the labor critical to Bengal's industries, but without the Government of Bengal's support, there was little hope for setting up a Pasteur Institute in Bengal.

Undaunted, the Government of India continued to pursue the establishment of medical laboratories and institutes. Between 1900 and 1914, Indian governments set up institutional structures and agencies that would shape the course of medical research in subsequent decades, effectively keeping medical research in the state sector.[68] By 1902, the production of anti-plague inoculations was in full swing, with the Government of India not only supplying Indian demands for the inoculation, but also sending tens of thousands of doses manufactured in the Plague Research Laboratory in Bombay to other British possessions, including New South Wales and Zanzibar.[69] The King Institute was built in the Madras Presidency in 1903 and the Central Research Institute in Kasauli in 1905. With an expanded research scheme in place, officers no longer needed to travel to Europe to study the bacteriology and parasitology of tropical diseases.[70] In fact, doctors from Goa, Germany, the Philippines, and other countries would later request permission from the Government of India to visit India's research laboratories.

Medical research in India, particularly the work on plague, had the intended effect in international circles. In 1909, the Indian Plague Commission had reported several landmark findings, namely that human–human transmission did not play a significant role in the propagation of plague in India, except in extremely rare cases of pneumonic plague. The Commission demonstrated that human infection depended entirely on infection of the rat population and that rat fleas were the vector of transmission from rats to humans.[71] The OIHP published these findings, which effectively conferred international approval on the commission's results and gave the Government of India an officially recognized scientific basis for its plague policies.

In its continued efforts to make research in India rank among the best in the world, the Government of India also established the Indian Research Fund Association (IRFA) in 1911. The IRFA concerned itself with researching the causation, epidemiology, and prevention of disease, especially of communicable diseases.[72] The Government of India wanted to convey an 'impression of the great discoveries which stand to her credit and the marked advance which she is now making in Public Health and Sanitary Science'.[73] At the first All-India Sanitary Conference in 1912, the president of the conference, Spencer Harcourt Butler opened the proceedings by emphasizing India's need for self-sufficiency in public health: 'We have to work out our own sanitary salvation. We have to study the epidemiology and endemiology of our communicable diseases, the so-called "tropical diseases" – plague, malaria, cholera and dysentery' in order to apply the best methods to prevent and suppress disease.[74]

Despite government avowals as to the critical need for research on India's diseases and the Government of India's recognition in the international public health community, India's progress in medical research had failed to reach the government's goals by the outbreak of World War I. Debates over the merits of different plague and cholera control strategies (discussed later in this chapter), financial and administrative limitations, and the difficulties of obtaining adequately trained personnel hampered the fulfillment of a program for a more adequate scientific study of the etiology and epidemiology of diseases in India. Progress in medical research in India had been hindered by inadequate staffing in spite of the development of new laboratories and the creation of a Bacteriological Department. With the onset of war, research took a further hit in India. IRFA activities and research were reduced: most research workers were funneled into the military, while the ones remaining in civil employ devoted their time to meeting war demands for vaccines and sera.[75]

After World War I, research activity increased and organization improved due to renewed efforts by the IRFA, but staffing levels remained low in the permanent cadre of researchers, and obtaining suitable men for available appointments proved difficult because of decreased recruiting for the Indian Medical Service.[76] Furthermore, the 1919 Government of India Act had sacrificed centralized science in an effort to promote political reform. Provincialization resulted in a weakening of India-wide scientific initiatives and research institutions as they became more subject to provincial financial and political pressures.[77]

The effects of provincialization were partly mitigated by the establishment of the League of Nations Health Organisation in 1923, which had the effect of boosting the Government of India's commitment to research. It provided a new forum within which India could establish its scientific legitimacy and be seen as part of a worldwide solution to prevent pandemics.[78] To achieve the Government of India's objective, Indian public health authorities sought both to align their research with the aims of this foremost international public health body and to influence international recommendations on disease control by serving on League committees. In the fall of 1927, the Director of the League's Far Eastern Epidemiological Bureau at Singapore recommended that the League's expert commission on plague include in its seven members two from India. This commission met in Calcutta during the Far Eastern Association of Tropical Medicine congress to draft a study program and coordinate international cooperation in plague inquiries.[79] The League circulated the report to all interested health administrations, suggesting that they model their plague research programs on the League's recommendations.[80] By 1930 the research program of the Haffkine Institute in Bombay had been aligned with the lines of research recommended at the Calcutta meeting of the League of Nations Plague Commission.[81] Then, in 1933, the OIHP asked the Government of India to begin investigations into particular aspects of cholera epidemicity and endemicity, focusing on the problem of healthy carriers. The OIHP's proposals for investigations were subsequently incorporated into the research agenda of the All-India Institute of Hygiene and Public Health, which had been established in Calcutta in 1932.[82]

Over the course of four decades, the Government of India had gradually achieved its goal of establishing India's pre-eminence in the world of medical research. The international community's concerns about the spread of plague and cholera had given the Government of India a compelling reason to actively pursue a policy of research. Medical research provided the Government of India with a means of countering the international community's negative perception of India's public health establishment. India's participation in international health forums and research reinforced India's place as an integral member of the international public health community. This created a vision internationally of a responsible India, which could only have helped mitigate any future political or economic risk that could arise with another disease outbreak. Political motivation was not necessarily a bad thing; the Government of India's support of plague and cholera research benefited both the reputation of India's public health establishment and Indians as well, or at least some of them.

## Endemic disease

### Dissent from within

No such prestige or trade politics would galvanize the Government of India to control endemic disease in the same way that it had for epidemic outbreaks. European states were not particularly concerned with India's morbidity and mortality rates, as long as disease remained within India. Mercantile organizations pressured the Government of India when trade embargoes threatened, but when the health of their plantation and mining laborers was at stake, they realized that the onus of disease prevention and treatment would be on them. By 1906, although plague mortality had risen, there was less government urgency to contain the disease, partly because health officials had begun to view the disease as one of filth and poverty and as one of the country's many endemic diseases.[83] The Government of India wanted to manage endemic foci well enough to keep the diseases local. How to accomplish that proved a contentious question.

Public health experts disagreed about the relative value of traditional plague control measures, including segregation, evacuation, and disinfection. Dr James Lowson, who had had experience in controlling plague during the Hong Kong epidemic and who had been Joint Plague Commissioner with the Government of Bombay, argued that home segregation was impracticable and ineffective and had led to much of the mortality in the early stages of the epidemic in Bombay, due to a failure to isolate all sources of infection.[84] Surgeon-Major-General Harvey, Officiating Director General of the Indian Medical Service, concluded that authorities should abandon compulsory isolation and segregation, except maybe in small towns or at the beginning of an outbreak when it could work.[85] Nield Cook, the Health Officer of Calcutta, also considered notification and segregation ineffective in a large 'oriental city' where the population objected to these measures. William Clemesha argued that early evacuation was the only useful and practicable measure for rural India but impractical in towns with a population above 10,000. Evacuation had drawbacks as well, particularly during the monsoons, and it disrupted business and schools and entailed significant expense. Based on his experience with plague control, Clemesha felt 'perfectly certain' that disinfection was futile and that in India, no amount of the most careful disinfection during a plague epidemic had had any effect in preventing a recrudescence of plague after the lapse of a quiescent or latent period.[86]

Instead of traditional measures, Cook advocated Waldemar Haffkine's vaccine 'of proved efficacy', which had been shown to reduce plague mortality by 80–90 percent among those inoculated. Considerable evidence also showed that the protection afforded lasted through at least one outbreak and potentially even longer.[87] Clemesha very practically pointed out that most natives, however, were still not inclined to undergo inoculation.[88]

Nevertheless, inoculation received further endorsement because of a widespread belief in the sanitary incorrigibility of Indians, which effectively circumscribed the types of disease control interventions officials were inclined to employ. Medical officers often located their explanations for endemicity in the domestic and religious habits of the natives, including insanitary usage of water, such as washing corpses and the deceased's clothes on the banks of rivers.[89] These were some of the many customs and practices that health officials abhorred and were unable or unwilling to try to change. Clemesha, like other medical officers, favored improved sanitation in theory but acknowledged that it seemed an impossible strategy in India.[90] J. A. Turner, the Executive Health Officer of Bombay was of the opinion that, 'Indifference and carelessness on the part of the poor people to accept any advice adds to the difficulties of control, while the ignorance of any measures of personal hygiene and the rooted objection to any innovation in their domestic arrangements form obstacles difficult to overcome.'[91] He believed that the habits of Indians, particularly the tendency to live in close proximity to animals, made India a 'country admirably adapted for the multiplication of rats'.[92] The government could not ignore the fact that many Indians, as believers of non-violence to animals, were unwilling to destroy rats as vehicles of plague.[93]

Official assumptions of Indian sanitary incompetence or incomprehension did not always go unchallenged. W. G. King, Sanitary Commissioner of the Madras Presidency, criticized officials for not enquiring as to whether these supposed Indian prejudices actually clashed with deep-seated religious beliefs. A policy of blame saved money and labor. Even when sanitary improvement schemes were approved, the government claimed lack of funds. On the contrary, King believed that funds were available, but sanitation was simply not given priority.[94] For officials, the beauty of inoculation was that it could overcome entrenched insanitary habits by bypassing them altogether. It did not depend on the individual's adherence to or rejection of certain practices and lifestyles. From the public health official's perspective, inoculation offered a more effective and practical approach.

While some medical officers approved of Haffkine's inoculations and his early data on efficacy, the Government of India initially hesitated to lend its support to these technological approaches to treating disease because it worried that protesters might riot.[95] The Government of India, therefore, remained unwilling to accept Haffkine's reports of the plague vaccine's efficacy in reducing mortality, despite its employment of him. Haffkine sent a report to the Government of India in June 1897 in which he argued that preventive inoculation was more effective than other measures. Segregation often led to flight and forcible removal, risking disease spread to non-infected areas. Inoculation involved no disruption to daily life and did not interfere with communication or traffic or require families to be segregated along with their sick. In May Harvey also issued a report denouncing the practice of isolation and segregation, which had failed except in isolated instances and had only led to concealment and dispersing of potentially infected people.

The Government of India stood firm in its opinion and sent a letter to provincial governments stating that it was not prepared to accept Harvey's conclusions. With the Bombay and Pune protests still fresh in the government's memory, officials were of no mind to introduce a measure, which could incite the masses. The IMS's official stance was that the only practical method of preventing and controlling the development of plague was to evacuate houses in which plague had appeared.[96] Clemesha argued that sanitation remained India's best hope to combat plague, especially in cities. Health authorities believed that the problem lay in the domestic habits thought favorable to the propagation of plague in India: overcrowded, ill-ventilated housing; lack of flooring in huts which enabled rats to enter dwellings; sleeping on bare ground; walking barefoot; and lack of knowledge or willingness to take early preventive measures.[97] Officials believed, however, that the standard of sanitation needed to alter those conditions would require impracticable changes in the dwellings and habits of ordinary Indians, effectively placing this method 'beyond the range of practical policy in India generally'.[98] Instead, measures of temporary utility, such as the destruction of rats, evacuation, and inoculation could be implemented immediately and result in large savings of life.[99] The difficulty of eliminating rats across the country led to a focus on removing them from commercial and residential properties.[100] Experts realized that measures of permanent utility, which focused on rat *exclusion* (for example, properly constructed dwellings, sanitation, depositories for food refuse, maintenance of grain stores and stables far from dwellings), involved enduring reforms, which, while necessary, were significantly more difficult.[101]

## Inoculation saves the day

The Government of India was at a public health impasse. Improved sanitation and rat elimination were impracticable, disinfection did not work, segregation and evacuation were resisted, and inoculation was unpopular. So what was the Government of India to do? The Indian Plague Commission resolved this dilemma in 1900 when it came out in favor of Haffkine's plague inoculations. The Government of India, satisfied with the Commission's evidence and lacking other options to address endemic plague, put aside its doubts and threw its support behind an inoculation strategy.[102] By 1907, inoculation had indeed become the central and provincial governments' anti-plague measure of choice.[103] Inoculation policy operated on a different basis of risk reduction and population protection than had previous isolationist or sanitary infectious disease control policies. It removed the necessity of individuals adopting new habits, but for inoculation to be useful, widespread adoption in a community was necessary.[104]

This technological approach to plague (and also cholera) was supported by a particular ideology emanating out of the London School of Tropical Medicine. The London and Liverpool schools of tropical medicine had been established within one year of each other at the turn of the century, yet they supported differing visions of imperial responsibility and represented different approaches to colonial development. The Liverpool school reflected the more practical interests of its merchant patrons and concentrated its efforts on investigating and improving sanitary conditions in the tropics. The London school's approach, which more heavily influenced colonial medical policies, focused on disease-specific research and individual prevention and treatment.[105] Its strategy of concentrating disease prevention efforts on the individual aligned well with the Government of India's inoculation policies.

Several persuasive tactics were tried in an attempt to make anti-plague inoculation more palatable to the masses. Cook had earlier suggested passing strict ordinances for isolation and segregation, so that natives would more readily submit to inoculation. He suggested giving a small monetary compensation to Indians as a further incentive. The Government of Bombay adopted his first idea to promote inoculation and began allowing inoculation to substitute for compulsory segregation.[106] Other provincial and local governments followed suit and sought to increase the popularity of anti-plague inoculation by exempting inoculees from more distressing measures. Lord Minto hoped that this method of persuasion would gain the assistance of the populace and allow governments to abandon more expensive and harassing measures. The Governor of Bombay, George Sydenham Clarke, also urged Indian language papers to support universal inoculation.[107] To further encourage inoculation in the presidencies, district collectors granted subsistence allowances, for a maximum of three days, to adults and minors temporarily incapacitated for work due to the inoculation. Although inoculation was officially voluntary in the Bombay and Madras presidencies, in cases where it was necessary to prevent plague, inoculations became essentially involuntary as officials invoked the Plague Regulations to threaten evacuation of the uninoculated.[108]

The central and provincial governments' efforts to promote inoculation worked. Despite early objections to inoculation and official perceptions of Indian fatalism, Indians adopted anti-plague inoculations fairly quickly, most likely because of perceived efficacy rather than government propaganda.[109] By 1916, the government laboratory in Bombay was producing upwards of one million doses of anti-plague inoculations annually. Demand for inoculation had increased to the point where in Vizagapatam in the Madras Presidency, the collector had to call in the police to prevent people from mobbing inoculators.[110] In 1935, the ratio of inoculations to plague deaths in the Bombay presidency had increased significantly from 9.1:1 in 1928 to 35.7:1 by 1935.[111]

Despite the success of inoculation and the Government of India's continued support for inoculation as the primary component of its plague policy, some public health authorities repeatedly insisted that sanitation and rats should be the primary targets. At the All-India Sanitary Conferences in 1912 and 1914, participants emphasized that, 'Plague preventive measures which aim at controlling human beings alone are insufficient and . . . it is more important to carry out those measures of permanent utility which tend to lessen the rat infestation of the house'. Proponents of a sanitation-centered policy advocated rat-control measures such as improved scavenging, bans on keeping livestock and grain storage facilities close to human dwellings, protection of grain stores, and increased attention to localities affected late in the season, which often became the source of early epidemics in subsequent outbreaks.[112] Systematic rat destruction was the best means of controlling and preventing plague.[113] Ironically, no laws existed to allow sanitary authorities to prevent people from harboring rats on their premises or to compel them to rat-proof their property and grain storage facilities. The director of the Bombay Bacteriological Laboratory, W. Glen Liston, believed this to be one of the weakest points in the Government of India's plague policy.[114] The Government of India countered that general rat destruction had a limited sphere of its utility; experience had shown that it resulted in only temporary decreases in rat infestation due to rats' 'excessive fecundity'. Of all the preventative options, the Government of India remained most optimistic about inoculation; it appeared efficacious, especially during epidemic outbreaks, and prejudice against it seemed to be diminishing in certain areas of India. The Government of India urged provincial governments to facilitate inoculation for anyone desiring it, but anti-plague inoculations would remain almost exclusively limited to towns.[115]

Despite the increase in inoculations, India was still the main reservoir of plague in 1925, supplying 90 percent of the world's cases.[116] With the development of bacteriology, health officials had focused on distributing and refining disease-specific technologies that could be employed at the individual level, resulting in limited measures to combat endemic plague and cholera. Furthermore, the central government had with the 1919 Act essentially transferred responsibility for public health and sanitation, including the containment of plague, to provincial governors with the aid of Indian Ministers responsible to provincial legislatures.[117] With the Government of India relinquishing responsibility for plague control, it was up to provincial governments to implement a progressive policy that would attack the sources of disease.[118] While the central government occasionally issued grants to the provinces for anti-plague measures and plague research, it had given up its role as a central coordinator of plague policy.

International health authorities recognized that India's plague continued to pose a public health threat, even though plague incidence was on the decline, and there was evidence of a decrease in virulence. But if plague incidence was not kept in check, it could easily infect port cities, risking spread to other parts of the world through maritime traffic. The League of Nations viewed plague as undoubtedly the most important Indian health problem for countries in maritime trade with India.[119] The League's viewpoint underscored the significance of plague in India for *other* countries. As long as the Government of India kept its ports free of plague, the League was less interested in India's interior. F. Norman White, Sanitary Commissioner with the Government of India, understood that the government needed to remain vigilant over all of India, not just the ports. He ascribed the change in virulence to increased rat immunity to plague and stated that it could in 'no wise be attributed to increasing efficiency of our ante-plague measures'. In the absence of plague, however, rat immunity was likely to decrease, which could lead to a consequent resurgence in plague.[120]

J. D. Graham, Public Health Commissioner with the Government of India, also warned against complacency, but more with an eye to the place of India in the international public health community:

> With the reawakening of interest in all social problems and the increased international activity regarding public health and the prevention of disease which occurred after the Great War, it was evident that much in the realm of national disease which formerly passed unnoticed must now come into the lime light and be subjected to international scrutiny. It was not enough for India to say that she was concerned with 300 millions of her own subjects and had no time to worry over the activities of other nations. She must be prepared to recognise that, in virtue of her important commercial position, she is an international offender – and a dangerous one as well – and in this spirit, to set about tackling the problem which confronts her by employing an organisation capable of utilising for such purposes the most recent discoveries of research in regard to these diseases and disease problems which are peculiarly her own.[121]

Despite official recognition of India's responsibility to gain control over its infectious diseases, both plague and cholera continued to haunt India. The heavy mortality toll caused by plague remained high in the 1920s with over 360,000 plague deaths reported in 1924.[122]

India had not fared much better in its control of endemic or epidemic cholera. It held the undesirable position of being the only country within the League of Nations in which cholera still existed in epidemic form. Annual mortality from cholera in the 1920s ranged from 67,000 to over 235,000.[123] Endemic foci of cholera existed in other countries, but India, and Bengal in particular, was widely regarded as the endemic home of cholera. Because of India's geographic and commercial position, this view was practically, if not scientifically, true. Thus international health officials looked to India to institute measures that would help rid the world of this 'very expensive incubus'.[124] As long as endemic cholera remained a widespread problem in India, the possibility of epidemic outbreaks would continue to loom.

The Government of India had placed itself in a constrained position with the 1919 act devolving public health to the provinces. It was neither constitutionally capable of forcing provincial governments to undertake sanitary improvements nor financially, administratively, or politically able to impose mandatory vaccination against cholera. Waste disposal and general sanitation were critical to controlling cholera incidence, yet sanitation was the affair of local public health departments.[125]

The only strategy to combat endemic cholera that the Government of India had left itself was further investigation of the disease at Indian research institutions and the manufacture and issue of anti-cholera vaccines.[126] Fortunately, the anti-cholera inoculation, like the anti-plague inoculation, had become increasingly popular. During a cholera outbreak in 1928 in Bengal, so many people rushed to receive the anti-cholera inoculation in Calcutta that the vaccine supply ran out after about 700,000 doses of vaccine had been administered.[127] In the 1930s, 1–1.5 million inoculations were being given annually in Bengal to prevent the disease spreading beyond the province's borders.[128]

Focus on cholera research and inoculations had disadvantages however. According to W. G. King, the Director of Public Health of the Madras Presidency, the development of new medical technologies for the prevention of cholera, including oral bilivaccines, bacteriophage, and water chlorination, had hindered public health progress in cholera-affected districts. Whereas cholera had previously acted as an incentive to improve elementary sanitation, which was the only means of ensuring radical reduction in cholera incidence, the disease could increasingly be dealt with on a more technologically sophisticated, if less effective, level. Medical technologies could decrease the incidence and severity of disease but did little to address the underlying causes of insanitary water supplies and poor hygienic practices.[129] Financial considerations continued to prevent or delay provincial governments' provision of clean water, especially to rural populations. F. P. Mackie, the Officiating Public Health Commissioner, noted:

> In one province the attention of the Board of Health was invited to a recommendation that at least one pakka well for the supply of drinking water should be built in each village of the province. It was found that with the grant at present available to the Board for this purpose a period of six hundred years must elapse before the programme suggested could be completed. Gigantic conceptions of this nature appear unlikely to be consummated until a demand for safe drinking water is created among the people.[130]

Furthermore, as had been the case with plague, many officials believed that the 'conservative and regrettably unhygienic customs of the people [would not] yield immediately to the influx of new ideas. Real progress [could] only be made along the slow lines of health education and improved water-supplies, food supplies, and improved systems for nightsoil disposal'.[131] This attitude stemmed from the ideologies of social Darwinism and scientific racism prevalent in the late nineteenth and early twentieth centuries, which allowed officials to rationalize colonization and public policies as natural consequences of the colonized's inferior culture.[132]

This pessimism pervaded the public health establishment and diverted efforts away from improving sanitation, but assumptions about Indians' inability to adopt the hygienic standards and practices were selectively applied. Otherwise, how could officials explain Indians' receptivity to education and improvements in maternal and child health and welfare or to other campaigns during the annual National Health and Baby Week held throughout the Madras Presidency, especially when the health week focused on promoting sanitary hygiene and disease prevention. In the 1932 campaign, cholera formed the focus of the Chingleput district health week, tuberculosis that of the Malabar and South Kanara districts, and anti-malarial measures for the Salem and Kistna districts. During the 1934 Health Week in Tanuku in the West Godavari district, people paraded through town a life-size picture of the goddess of health. The watercolor picture depicted the goddess with a crowned head and ten hands, each carrying different instruments related to preventive measures, including a lancet, hypodermic syringe, scalpel, quinine, and cholera tablets.[133] These Health and Baby Weeks were successful because they involved cooperation among multiple parties: provincial government, local boards, municipal councils, local health associations, and the Indian Red Cross Society. Clearly, some officials and health experts believed that health education could 'arouse a sense of civic responsibility' and create a desire for self-help within each person.[134] Provincial governments utilized propaganda on the causation, mode of transmission, and prevention of cholera in addition to chlorinating water-supplies, disinfecting infected homes, and providing anti-cholera inoculations, isolation, and treatment.[135] Despite some central funding of public health campaigns, negative perceptions of Indians' capacity for sanitary improvement persisted at the federal level well into the 1930s and continued to impede government-led health progress.

Indian critics asked why, if Indians had such bad sanitary habits, did the British not undertake widespread instruction on hygiene.[136] In the late 1930s, nationalist N. N. Gangulee, who had also served on the Royal Commission on Indian Agriculture from 1926–8, critiqued British policy of non-interference with Indian beliefs and customs, which then resulted in government neglect of public health. He acknowledged that

Indians were often resistant to Western medicine and sanitation but maintained that the Government of India still had the responsibility to educate them.[137] He also argued that health authorities did not address the association between plague incidence and poverty in India, either in lowering resistance or in precluding sanitary and well-constructed dwellings. Europe had not eradicated the rat to eliminate plague; it had improved the standard of diet and habitation. He pointed out that unless people could afford better and more food and build sleeping quarters off the ground and away from rats, no effective control of plague could occur.[138] But addressing the roots of poverty was beyond the political will of the colonial government, whose officials would continue to promote inoculation as an easier and more economical method of control that could quickly and directly reach millions of Indians a year.

## Conclusion

By 1940, the mortality rate for plague and cholera had fortuitously fallen compared to that of the 1890s and early 1900s.[139] A variety of factors had helped lower plague mortality: acquired rat immunity, the cyclical nature of epidemics, millions of anti-plague inoculations, and efforts to reduce the proximity of rats and humans.[140] Cholera mortality had also declined due to an increase in inoculations, sanitary vigilance at religious festivals, and improvements, mostly in urban areas, in water supplies and waste disposal practices. Nevertheless, the number of deaths remained high, with plague claiming tens of thousands of lives annually and cholera up to 240,000 lives yearly in the 1930s. Cholera was no longer as serious a problem in the larger municipalities, but it remained of profound importance throughout India as an epidemic scourge.[141] With both cholera and plague, a recrudescence could not be ruled out.

Despite the continued high mortality rates from these diseases, the Government of India's plague and cholera policies were inconsistent and its actions often sporadic and confined to certain places or groups of people. These policies resulted from the Government of India's perceptions about the consequences of the two epidemiological forms of disease – the epidemic and the endemic. These two forms provoked different reactions within and without India, reactions which then pushed policy in certain directions.

Responses to epidemic cholera and plague resulted more from impending political and economic crisis and reactions to European anxieties than from concerns for public health. Epidemic outbreaks of cholera and plague in India raised alarm in Europe and in the Ottoman Empire. After 1890, these fears coincided with the growth of an increasingly strong international public health community, which could force India either to deal effectively with epidemics or to face deleterious consequences to its trade. Epidemic outbreaks also placed India's international trade, domestic economy, and political relationships at risk. International reproach and trade embargoes and the looming threat of future trade restrictions combined with increasing pressure from India's mercantile associations spurred the Government of India into action. In order to protect India's economy, the government had to contain epidemics and prevent endemic pockets of plague and cholera from ballooning into epidemics. Epidemic outbreaks, thus, brought out the colonial government's arsenal of responses: medical observation of travelers; evacuation; forced hospitalization; disinfection; control of ingress and egress from infected areas; inoculations; rat destruction; and regulation of grain traffic. Because of the epidemiological connection between large mass movements and the spread of cholera and (as was initially believed) the spread of plague, the regulation of both domestic and international pilgrimage, and hence the control of pilgrim movements, personal effects, and bodies, also became a critical part of the Government of India's crisis-driven plague and cholera policies. Trade restrictions had been the initial impetus behind government action, but pilgrimage bans were the effect.

The Sanitary Commissioner with the Government of India wrote in his 1900 Annual Report that 'one of the outstanding mysteries of our policy in regard to plague in the eyes of intelligent foreigners is the fact that the epidemic is treated rather as a political emergency than a matter of public health'. The public health legal code in India supported this shortsighted approach to disease control.[142] The Epidemic Diseases Act, arguably the most important act of public health legislation passed before the Devolution of 1919, aimed first and foremost to prevent the spread of epidemic disease from India to other countries. In formulating its legal and medical strategy to fight epidemic plague and cholera, the Government of India primarily wanted to ensure compliance with the international sanitary conventions, protect trade, and assuage any fears abroad of the potential spread of plague or cholera outside India. Because of the need to maintain international confidence in India's efforts to prevent the spread of epidemics, the Government of India often went beyond international protocol and implemented standards of inspection and disinfection at ports for pilgrims and sailing vessels that were stricter than those stipulated by the 1897 Venice convention.[143]

While appeasing European states, the Government of India also needed to accommodate domestic priorities. If India was to remain politically manageable and economically viable, the government could not forever ban Indian pilgrimage to the Hedjaz just as it could not permanently restrict labor movements. Maintaining Muslim support within India had underpinned the Government of India's stance toward minimizing pilgrimage restrictions, just as the protection of the colonial economy had underlain the Government of India's decision to exempt laborers from certain disease regulations.

In the end, the Government of India achieved its goal of protecting the health of Indian trade and its international and imperial political relationships. Improved understandings of disease etiology had led to better methods of containment. Its 'vexatious and irritating' plague measures, central and provincial governments' restrictions on travel and pilgrimage, and the implementation of the Venice Convention had enabled Indian trade to communicate freely with European ports with only minor interruptions.[144] Research being conducted in India had tangibly assured Europe of India's good faith efforts to tackle its diseases and moved India into a place of prominence within international public health circles.

In understanding the Government of India's risk behavior, it is necessary to recognize that different societal actors, such as government bureaucracies, trade lobbies, and the medical profession, influenced risk characterization and risk management strategies and were critical actors in defining acceptable risk. Organizational interests shaped qualitative estimates of risk and created the potential for conflict among various political and economic constituencies involved in risk management.[145] These conflicts were resolved by the Government of India's prioritization of its economic and political relationships with Europe and both parties' desire to keep those relations mutually beneficial.

No such imperative existed for endemic disease however, and thus the Government of India never felt the same sense of urgency toward endemic plague and cholera. Political, economic, and health risk in relation to endemic disease was not characterized by officials in India or internationally in the same way that risk from epidemic disease was. Deaths resulting from endemic disease had fewer, if any, international economic or political ramifications. No perceived mutual benefit for Europe and India existed to compel the Government of India to consider endemic cholera and plague comprehensively.

Endemic infectious disease policy followed the path of 'least financial resistance'. The Government of India's public health expenditures as a percent of total expenditures declined from a high of 1.3 percent in 1890–1 to a low of 0.3 percent in 1938–9. The Government of India's allotment to the provinces for public health expenditures following Devolution did not make up for this decline in central health expenditure either.[146] Few provincial governments were able to finance the practical application of research and knowledge, especially during economic downturns, which resulted in a failure to profit from the knowledge at India's disposal. Public health was often the first to get pushed aside when it came to cutting budgets.[147] Proper sanitation was considered financially and logistically difficult, leaving inoculation as the measure more in line with colonial public health budgets. Provincial governments found it impossible to introduce permanent plague-control measures in rural India and so focused attention on a limited number of endemic areas of plague while continuing to encourage inoculation.[148] Budgetary restraints meant that strategies to reduce cholera and plague incidence in the long run would never form a part of central policy in colonial India, while the institution of Devolution in 1922 would leave provincial and local governments to decide policies based on their particular priorities and financial and administrative capabilities.[149]

Assumptions about Indian norms and practices and opinions of Indians as a hygienic other also pushed policy toward inoculation. Medical thought and practice, and, therefore, also policy, were constrained by cultural and intellectual institutions and perceptions of disease and peoples.[150] An essentializing, Orientalist discourse about Indian cultures and societies eliminated individual Indian agency as a public health strategy, justified colonial officials' attitudes toward the feasibility of certain public health policies, and rationalized government inaction.[151] The perception of a static India permeated realms of colonial thought into the twentieth century and affected how public health officials perceived the place or inevitability of epidemic and endemic diseases in India's health landscape. Despite colonial rhetoric of the civilizing mission, the state was often constrained by its bureaucracy and fatalism – why waste money on health and sanitation for an unreceptive population? Many officials believed that education was the only means to win Indians over to Western medicine and sanitary practices, but they recognized that this would require substantial amounts of time and money.[152]

Inoculation had perpetuated the Government of India's dichotomous approach toward epidemic and endemic disease by allowing it to address endemic disease superficially without having to launch an attack on the underlying causes of plague and cholera. On the other hand, epidemic outbreaks, because of the economic and political risks they carried, forced the Government of India to address these diseases on both levels. Cholera and plague continued into the 1940s to present significant health problems to Indians, but they no longer posed the same epidemiological risk to European and other states as they had in the late 1890s and early 1900s.[153]

With plague and cholera in India receiving less international attention by the outbreak of World War I, the Government of India could focus more of its efforts on addressing malaria, which in most years, killed more Indians than plague and cholera combined. While the underlying motivations of malaria policy differed in many instances from plague and cholera policies, they paralleled each other in practice, at least where endemic disease was concerned. Central policies regarding endemic malaria and endemic cholera and plague privileged medical technologies and medicines over sanitary progress or elimination of the disease vector. Malaria policy focused on distributing quinine and researching quinine alternatives, just as cholera and plague policies concentrated on distributing inoculations. However, unlike epidemic cholera and plague, epidemic malaria did not threaten to spread to other countries, nor did it affect international trade. These fundamental differences lay at the core of the Government of India's approach to malaria.

# 3
# Malaria – India's True Plague

> *In this, O Nature, yield I pray to me. I pace and pace, and think and think, and take The fever'd hands, and note down all I see, That some dim distant light may haply break.*
>
> *The painful faces ask, can we not cure? We answer, No, not yet; we seek the laws. O God, reveal thro' all this thing obscure The unseen, small, but million-murdering cause.*[1]
>
> Ronald Ross

So wrote a melancholy Ronald Ross as he toiled in India, hoping to discover the means by which to cure this dreadful scourge. In 1898, his years of research in outposts of the Indian Medical Service would finally be validated when he discovered that the malaria parasite was transmitted through mosquito saliva during the act of biting. Although Ross' malaria research was indisputably a monumental advance in understanding malaria's epidemiology, the implications of his research for India were far from clear.

Malaria killed and caused more sickness and death than any other disease in India.[2] Throughout the late 1800s and first half of the 1900s, officials estimated that malaria afflicted 100 million Indians annually and killed over one million per year in India alone.[3] The disease inflicted an enormous toll on Indian health, as did the debility caused by the lasting effects of malarial illness. Malaria workers saw that malaria increased sufferers' susceptibility to chronic and other acute infections, especially diarrhea, dysentery, pneumonia, abscesses, kidney disease, anemia, and convulsions in infants. Furthermore, maternal morbidity and mortality due to malaria significantly increased the child death rate. In 1931 the disease killed directly and indirectly 2,172,214 people, while the morbidity from malaria-related causes was estimated to be as high as 75 million cases annually.[4]

Malaria in India was a complex problem with no simple solution, but over the course of the first half of the 1900s, investigations and mosquito surveys conducted in India helped refine malariologists' understandings of the particularities of India's malaria epidemiology.[5] Public health officials understood that malaria was primarily a rural disease, with a marked difference in malaria incidence between rural and urban parts of India. The Government of India estimated that 900,000–980,000 of annual malaria deaths occurred in rural areas, where treatment was scarce, funds even scarcer, and sanitation poor.[6] The malaria death rate in villages was estimated to be double of that in towns. Public health officials believed that urban areas were much less affected by malaria and urban case mortality significantly lower due to a greater availability of treatment centers in towns and fewer breeding sites for mosquitoes.

Part of the peculiarity of India's malaria situation was that much of India's malaria, experts believed, had been man-made. The drainage problems created by construction works had increased the population of malaria-carrying mosquitoes. The concomitant recruitment of laborers for these works had introduced human malaria-carriers into unaffected areas; conversely, economic opportunity had brought non-immune populations into malarious areas. In irrigation projects, which were largely managed to achieve greater profitability, engineers neglected drainage while concentrating attention on the more lucrative parts of irrigation, which included supplying the water requirements of crops and devising means to deliver water quickly and cheaply.

Development had brought in its wake immense agricultural disruptions and unanticipated repercussions for public health.[7] The construction of railways, roads, bridges, and canals, especially in western Bengal, had interfered with natural drainage and flooding. This in turn had caused bodies of water to dry up and had impoverished the soil, leading to less abundant harvests. Railway embankments and altered water flows also created accumulations of stagnant water in some areas and a subsequent increase in malarial breeding grounds.[8] Improper canal irrigation had set into motion a deleterious chain of events. It created waterlogging, which led to endemic malaria and consequent poor health; poor health led to economic stress, which resulted in greater privation and sickness, provoking a high death rate, low birth rate, emigration, and eventually depopulation of affected tracts of land.[9] Villagers deserted their lands in an effort to escape the devastations of malaria.

Desertion combined with high mortality levels led to a contraction in the area of land under cultivation. In various districts in west and north Bengal, cultivation diminished by as much as 60 percent between 1891 and 1931. Reduced agricultural capacity led to even greater poverty and insufficient nutrition, further exacerbating the incidence of malaria and creating a vicious cycle of illness and insufficient food supplies.[10]

Malaria had a devastating social impact. Repeated bouts of the disease hit the poorer classes of laborers, settled peasants, and artisans particularly hard.[11] The disease picked away at people, ripped apart social fabrics, and disrupted family bonds and structures when heads of households, wives, or children became disabled due to disease or died. In endemic areas, especially in malaria-ravaged Bengal, many had to resort to begging, reducing them to social outcastes.[12] Malaria decreased fertility and induced spontaneous abortion, thus depressing population growth, shortening life expectancies, and ravaging families.

Medical congresses in the 1930s also drew attention to the economic impact of the disease. Experts emphasized that high morbidity and mortality rates caused 'incalculable financial loss due to diminished efficiency, loss of industrial development, and failure to exploit agricultural and mineral resources'.[13] Malaria caused the government losses in two major ways: in revenues, the Government of India experienced a loss of land tax-returns and decreased returns on government profit ventures; on the expenditure side, the government incurred higher costs for public works designed to exploit natural resources, lost money in the forms of extra pay and medical expenditure to those stationed in malarious areas, and incurred the medical costs of providing quinine treatment in hospitals and dispensaries for therapeutic purposes.[14] The annual economic loss to India for direct malaria mortality, calculated based on the loss of revenue to the state and the economic value of a life, was estimated at £67.5 million; the financial loss to the individual or family annually about £80 million; and the economic loss to the community, including to business people and government, due directly to malaria sickness (and not indirectly to the retardation of economic and industrial development) was estimated at £26 million annually. Malaria threatened the strength and productivity of India's labor population and economy and the nation's overall health.[15] Given that the majority of India's exports were raw materials produced from labor-intensive agriculture (jute, cotton, rice, tea, oilseeds, wheat, raw hides), the debilitation of India's labor supply was even more critical.[16]

Despite the astounding economic burden of malaria on India, the total budget for *all* medical concerns of the nine major Indian provinces during the financial year 1931–2 was a paltry £3 million.[17] The total number of cases of malaria treated in India amounted to only 8,678,664 between 1925 and 1929. The prevalence of treatment did improve somewhat, so that by 1931, over 11 million cases were being treated annually in hospitals and dispensaries. However, a case did not necessarily equal a unique malaria sufferer. Officials assumed repeat cases, meaning that in reality, less than ten percent of the malaria-afflicted population actually received treatment.[18]

This recognition of the negative economic impact of malaria was not a phenomenon of the 1930s; it had attracted government attention as early as 1858, when the Government of India began cultivating cinchona trees to produce quinine.[19] Not until 1909, however, when the Government of India convened the Imperial Malaria Conference at Simla, did the government formally recognize that India needed a general malaria policy.[20] The most important resolution coming out of the Simla conference provided for the formation of a Provincial Malaria Committee in each province, which meant practically, that from 1909, active malaria control was largely left to the provinces, with the Government of India retaining responsibility primarily for quinine policies and malaria research.

Public health officials and politicians understood that malaria hindered not only the economic progress of the country, but also its social, intellectual, and political progress. They also knew how difficult a problem it was to address. The agricultural, economic, and sanitary conditions within India, the diversity of climatic environments, and the varying prevalence of several species of mosquito vectors made it clear to health officials and malaria experts that no one anti-malaria measure was suitable to all locales and no simple or surefire method existed to eliminate mosquito breeding grounds.[21] They realized that malaria was India's most urgent disease problem and health risk, which begs the questions: Why did the Government of India not prioritize public health and clinical interventions to combat malaria in the way it had for epidemic cholera and plague? Why into the 1930s did malaria policy remain incoherent and limited in scope? And why, 80 years after the government had begun growing cinchona for quinine, was malaria still devastating India's population?

## Quinine policy

Like other parts of the malaria-afflicted world, India relied heavily on the use of quinine as treatment. Among cholera, plague, malaria, and yellow fever, malaria was the only disease with drugs for its prevention and treatment and the only one for which an inoculation would not be developed during this period.[22] The Government of India began to depend more on quinine in the mid-nineteenth century, when the Indian and Dutch government policies of encouraging the private cultivation of cinchona plantations led to a drop in the price of quinine.[23] This, in addition to the development of a more efficient manufacturing process to produce pure sulphate of quinine at the Bengal cinchona factory, had by the early 1900s rendered practicable a scheme to distribute quinine to poor peasants in malarial areas. In 1892, the Government of India sanctioned a program for large-scale sale and supply of government-procured quinine to the poorer classes in the Bengal and Bombay provinces. This scheme had dual purposes: it endeavored to increase access to quinine for villagers living in rural areas far from dispensaries and to popularize the drug's use.[24] Between 1890 and 1894, the Government of India's 'pice packet' scheme, which entailed selling single doses of quinine through local post offices, was gradually introduced into various provinces.[25] Quinine quickly became the backbone of the Government of India's malaria policies, which aimed, firstly, at attacking the parasite in man through quinine treatment and prophylaxis and, only secondly, at eliminating mosquito breeding grounds.

The Government of India's pharmaceutical emphasis did not come without complications. Health officials early recognized the need for centralized control of quinine manufacture and the possibility of an impending quinine shortage. In 1895, the Government of Bengal pointed out that supplies of stored and harvested cinchona bark, from which quinine was derived, would be exhausted in ten years' time. Planting trees on the uncultivated portion of government land designated for cinchona farming would extend supplies by only three years, and up to that point, replanting trees had met with little success. The Government of Bengal urged the Government of India to assume responsibility for India's quinine supply, warning that it would be disastrous for India's public health if the Government of India did not exercise sufficient foresight to guard against exhausting all quinine stocks in the future.[26]

The perpetual inadequacy of quinine supplies and of the government's reserve stocks would haunt the Government of India. Reserve stocks were primarily intended as a hedge against serious epidemic outbreaks, market volatility, or a squeeze from Dutch Javanese cinchona suppliers, who supplied the bulk of the world's cinchona bark.[27] During epidemics, the Government of India would provide quinine without charge to hospitals and dispensaries and send hospital assistants to distribute the drug in malarious areas.[28] During normal malaria years and when the quinine market was stable, the Government of India held on to its reserves and continued to encourage manufacture and importation. By 1906, however, the continued rapid growth in quinine consumption had begun worrying the Government of India that demand would soon outstrip the increased supply resulting from improvements in manufacture.[29] But rather than increase manufacturing capacity or extend cinchona cultivation in India, the Government of India began to reconsider its policy of self-sufficient quinine production, since quinine could be bought on the open market as cheaply as the government could manufacture it.[30] Given that total Indian quinine consumption and production accounted for only a small fraction of global consumption and manufacture of quinine, the Government of India did not think that India's withdrawal from quinine production or even a doubling of its consumption would increase market prices.

The problem with the Government of India withdrawing or reducing its manufacture of quinine was that it put itself at the mercy of Javanese suppliers – a precarious proposition since India purchased nearly one-sixth of Java's annual cinchona bark supply. The dilemma lay in that Java's climate and geography were extremely conducive to cinchona growing, making it impossible for India to compete on cost or quantity. A large-scale extension of planting to provide India with enough cinchona bark to meet quinine demands was simply not financially or agriculturally practical.[31] To hedge against increasing Indian dependence on Java's cinchona supply, health authorities at the Simla Imperial Malaria Conference recommended extending cinchona cultivation to increase quinine output in India. This seemed prudent, since the maximum output of Indian factories was 100,000 pounds, sufficient to treat only 3,000,000 people, and since the government cinchona plantations plus a small reserve of manufactured quinine provided the only barrier to a complete Javanese quinine monopoly.[32] But since cinchona trees needed eight to nine years of growth before the bark could be harvested without damaging the trees, this solution to Java's quasi-monopoly would not bear fruit for at least another decade.

The Government of India realized the need to consider these limitations in its future quinine distribution policy, especially one that would extend the drug's distribution. Unless an adequate supply at a reasonable cost could be assured, the distribution system would break down.[33] Despite this consideration and the conference's suggestion, the Government of India still did not see an immediate need to expand the area under cultivation. It did decide to prospect carefully, so if extension became necessary, there would be no delay in planting.[34] This policy was shortsighted though, given the long time lag between planting and harvesting bark.

The quinine/cinchona dilemma would become even more complicated. In 1912, government officials speculated that Java plantations might not have sufficient capacity to continue to supply the world's demand, thus underscoring the importance of Indian cinchona plantations. The Government of India's quinine reserves were sufficient to treat a severe malaria epidemic but were not enough to provide for a considerable increase in future quinine consumption.[35] Furthermore, increased demand from countries in tropical and subtropical areas had resulted in the price of quinine jumping by over 25 percent in 1912. This led the Government of India to increase its quinine reserves to decrease its vulnerability to price fluctuations for quinine.[36] The government decided that these circumstances warranted expansion and granted the Government of Bengal funds to extend cinchona planting in Bengal.

In the interim, since Indian plantations were meant to serve primarily as a reserve source, arrangements were made to purchase large quantities of bark from Amsterdam and Java. To ensure the success of this policy, Indian government factories manufacturing quinine from purchased bark needed a guaranteed quinine demand, which necessitated that quinine be sold at a reasonable price to provincial governments. Provincial governments, hospitals, and dispensaries in turn were then supposed to buy all quinine from designated Indian government factories.[37] However, formulation of quinine policy at the center and its implementation in the provinces were disconnected. Jurisdictionally, the Government of India could not force provincial governments to buy quinine, and even if the provinces wanted to purchase the drug, not all provincial governments had the funds available to obtain quantities of quinine in proportion to their malaria needs.

The internal political and agricultural difficulties of procuring adequate amounts of quinine, or at least enough to meet demand, in conjunction with the Javanese monopoly presented huge impediments to the practical realization of the Government of India's quinine policy. The Government of India itself created another obstacle by showing greater concern for protecting private industry than for the ability of the poor to purchase quinine. Although there were two government factories designated specifically for the manufacture of quinine, one run by the Government of Bengal and the other by the Government of Madras, the Government of India stipulated that provincial governments should not manufacture or sell quinine in a manner that interfered with private enterprise. When the Bengal factory initially began manufacturing quinine, it eliminated the need for the Government of India to purchase quinine from the private sector, thus depriving private suppliers of one of their largest buyers of private quinine. Because government manufacture cut into private sector profits, the Government of India decided that as a means of compensation, government factories should abstain from open competition with private enterprise. Quinine made in Bengal and Madras was subsequently only issued to government medical depots, government officers, and medical missionaries. Quinine packets sold to the public through government institutions were priced at a pice because it was high enough to protect provincial governments from any criticism of underselling private manufacturers of quinine, but believed to be low enough to make it available to the very poor, although the poor's purchasing power would later be called into question.[38]

The Government of Bengal agreed with the Government of India's pricing scheme, even though it argued that the sale of pice packets to the general public did not really compete with private industry. The Government of Bengal believed that pice packets were being bought by a stratum of the population that would not normally buy quinine. Furthermore, the private sector did not have the extensive distribution system that the government had (via post offices) to even reach this stratum. But since finding suitable government land to extend cinchona cultivation had proven difficult, the Government of Bengal thought it critical to government stocks that private enterprise did not withdraw from quinine manufacture. To prevent a withdrawal, both the Bengal and Madras governments needed to avoid undercutting private sales.[39]

Reluctant to compete with private enterprise, the Government of India began encouraging a greater importation of quinine, which increased the amount of quinine available for private sale. Between 1900 and 1910, the amount of quinine imported into India grew 88 percent, from 63,812 pounds to 120,112 pounds. By contrast, the government manufactured only 45,864 pounds of quinine in 1910.[40] The Government of India's concern for the private quinine industry in India paralleled the protection of commercial interests that had factored into the decision-making process for the government's cholera and plague policies. Whereas the strict precautions put in place to protect overseas trade during plague and cholera outbreaks benefitted the health of Indians by containing epidemics, incentives to increase quinine imports, which were then sold at a price higher than governmental quinine, reduced Indians' access to treatment.

Despite the intricacies and difficulties of establishing a coherent, effective, and sustainable quinine program and the fact that the Government of India would never be able to obtain enough quinine to treat all of India's malaria sufferers, the focus of malaria control continued to be directed primarily at quinine prophylaxis and treatment. Public health officials ignored the Royal Society's recommendations to prioritize mosquito eradication. They felt justified in their approach, however errant in retrospect: they cited Angelo Celli of Rome, a world-renowned malariologist in Europe, who had found that mosquito destruction in extensive areas had 'generally insuperable' difficulties.[41]

The Government of India's quinine emphasis was buttressed by experiments in India as well, particularly by the failed two-year attempt to eliminate mosquitoes in the Lahore cantonment of Mian Mir between 1901 and 1903. Despite these results, Ronald Ross vigorously defended mosquito eradication. Ross had spent almost 20 years in India as a military doctor and researcher of tropical diseases. He argued that the Mian Mir experiment had been carried out incorrectly and with insufficient consideration of the methods, factors, and conditions affecting outcomes. Ross would have known; he was well-versed in the peculiarities of malaria in India. He had been awarded the Nobel Prize in 1902 for his practical and theoretical research on malaria and was arguably the foremost expert on malaria at the time.[42] Since he viewed the Government of India's focus on quinine as short-sighted and instead advocated a comprehensive malaria policy that attacked the disease on the parasite, vector, and host levels, Ross had become *persona non grata* to the Government of India. Ross wrote that Mian Mir was not a favorable place for mosquito reduction and lamented that the Mian Mir debacle had prejudiced public opinion and that of less knowledgeable authorities against the feasibility of mosquito reduction. The Government of India went on to publish the Mian Mir report and inquiry, which effectively constituted an official pronouncement that mosquito reduction was difficult, expensive, and not very efficacious. Ross believed that this position had damaged the future of malaria control in India.[43] The Government of India thus continued to keep quinine at the center of its anti-malaria policy. It relayed to provincial governments that the popularization of quinine must be the goal, at least until the provinces could undertake extensive drainage works. The Government of India told provincial governments to ensure that every household knew the value of quinine, how to take it, and where to get it.[44]

## Critiques of policies

Although the Government of India continued to focus on quinine and cinchona policy, it recognized the need to establish an organization, both provincial and central in nature, to study and prevent malaria in India. The severe malaria epidemics that had racked the United Provinces and the Punjab in 1908–9 gave further impetus to this idea. The estimated mortality rate at the time from malaria was approximately 1.13 million people annually. The severity of malaria in India made it imperative for the central public health establishment to reach a consensus on how to decrease the disease's incidence. To this end, the Government of India convened the Imperial Malaria Conference in Simla in 1909 to discuss India's anti-malaria strategy.[45] At the end of the conference, participants concluded that a first step in any malaria campaign was to obtain detailed knowledge of the circumstances associated with the disease in different localities, since malaria thrived in such diverse conditions throughout India.[46] They supported extensive scientific investigations into malaria's epidemiology and endemiology in India, the modes of action of quinine and other remedies, and cost-effective methods of extirpating mosquitoes. The conference advocated the formation of a central scientific committee to direct and coordinate studies, including research on quinine treatment and prophylaxis. They agreed that a designated sanitary staff should be specially trained to address malaria and that residents in malarious areas should be educated on malaria prevention.[47]

Following the conference, the Government of India did establish the Central Malaria Bureau to house a permanent laboratory for malarial investigation and teaching, but not all of the recommendations necessarily translated into policy, especially since public health officials sometimes held different views of the best strategy to pursue. Expert opinion for treating and preventing malaria in India divided into two camps.[48] The 'quinine party' was committed to general quinine distribution as the primary malaria policy and, according to Ross, consisted of influential tropical medicine specialists such as the Indian Medical Service's J. T. W. Leslie, S. P. James, and S. R. Christophers. The Government of India largely adopted the quinine party's stance and based its policies on the grounds (albeit contested) that quinine was the cheapest anti-malarial measure. Ross contended that the Simla conference, which had been organized by the 'quininists' and to which he had not been invited, was biased; neither military medical officers trying to eliminate malaria in cantonments nor engineers were invited to the conference.

Although most delegates were malaria experts and had succeeded in passing some resolutions that encompassed more than quinine treatment, Ross doubted that these resolutions would be put into practice, since the quinine party retained executive power.

In the second camp were the 'opportunists', who, like Ross, favored using any measure likely to achieve the best results given local conditions and cost. Ross, who belonged to an older tradition of holist, preventive, sanitary medicine embodied by the Liverpool School of Tropical Medicine as opposed to the newer curative and reductionist laboratory medicine practiced at the London School of Tropical Medicine, advocated the elimination of mosquito breeding grounds and the use of mosquito nets and wire gauze protection of houses, in addition to quinine prophylaxis.[49] Furthermore, controversy over the role of quinine in the etiology of blackwater fever also prejudiced some IMS officials against the use of quinine as a prophylactic and treatment for malaria.[50]

The two camps disagreed not only on overall approaches but also on the cost-efficacy of different interventions. J. T. W. Leslie, the Sanitary Commissioner with the Government of India, pointed out at the 1909 Simla conference that the 'continuous use of quinine, even for a short time during the year, is inconvenient and unpleasant to the individual and difficult to carry out among a community. It is therefore evident that the best way to get rid of malaria is to destroy the mosquitoes. The only questions are: can it be done? and, if it can, at what cost'?[51] The quininists contended that mosquito reduction was too costly to be feasible, while the opportunists thought a general quinine campaign would cost too much. Disagreement within the public health establishment hindered the development of a more comprehensive malaria policy, just as it had done with cholera and plague. In the end, the Government of India sided with the quinine camp and refrained from implementing a mosquito reduction policy, believing that it entailed enormous expenditures beyond the means of any Indian town, much less rural India. Ross believed that cost was the primary factor in the government's narrow approach to malaria control.[52]

Ross was not the only critic of the Government of India's malaria policy.[53] Patrick Hehir, member of the Indian Medical Service and a malaria specialist, faulted the Government of India and provincial governments for limiting expenditure on malaria to the amounts allotted in the medical and sanitary budgets, and during epidemics to the amount of special grants.[54] W. G. King, Director of Public Health for the Madras Presidency, viewed policies that assumed localities should pay for anti-malaria measures as preposterous given the inequitable distribution of funds among and within provinces. Local authorities did not have resources at their disposal to undertake drainage, install piped water supplies, or rectify faulty railway embankments; they sometimes even lacked adequate authority to enforce preventative measures, such as the elimination of stagnant water pools.[55] Hehir argued that the limit on malaria expenditures should equal the cost of malaria to India, including the costs of increased mortality, decreased production capacity, treatment of illness, and lost labor time. King labeled the Government of India's quinine policy 'cheap policy'. It placed the onus on the masses to buy quinine and cost the government practically nothing.[56]

Furthermore, a quinine-focused policy was injudicious. For 'India to trust to sole quinine prophylaxis – even were the effort confined to the rural population – is a mere dream and should never have been placed before a Government *as a prime policy*, when dealing with a country where at the mildest computation, there are annually one million deaths and over three hundred million cases of sickness with the attendant disturbance of the economical resources of the country to be thought of', not to mention the negative impact on the already small earnings of the average laborer.[57] Even if world quinine production tripled, there still would not be enough to treat or prevent malaria in India. As of 1913, consumption of quinine in India had reached one-sixth of the world's production, but this quantity was still far from sufficient for comprehensive prophylaxis or treatment. The Italian experience of treating malaria, which served as an oft-quoted model for India, showed that quinine should be distributed to laborers at public expense and that voluntary submission to prophylactic quinine was not to be expected. Charles Bentley, Sanitary Commissioner for the Government of Bengal, argued that all the means of distributing quinine to the populace up to that point had proven inefficacious. He advocated a major pro-quinine propaganda campaign, similar to the Italian model, to convince the population of the medicine's benefits and to stimulate demand and compliance.[58]

Quinine was sometimes distributed in India at no charge, but the policy still had dubious results. Dakshina Ghosh, Deputy Magistrate and Deputy Collector in the district of Birbhum in Bengal, questioned whether quinine was sufficient to reduce the malaria mortality rate. Deaths in his district remained high at 40 per thousand despite a widespread distribution of quinine and having good drainage. He noted that quinine was most efficacious as a prophylactic in earlier bouts of malaria, but when secondary diseases arose consequent to malaria infection, such as pneumonia, heart problems, and dysentery, other medicines were needed to treat the sick poor, or they would die by the thousands. The annual distribution of quinine did not produce lasting effects, and Bengal, like the rest of malarial India, needed a strategy of more localized intervention to achieve definitive improvement in malaria incidence.[59]

Experts also critiqued the Government of India's inadequate support for educating Indians about the importance of malaria control, proper sanitation, and methods for malaria prevention. Some malaria experts thought it peripheral to an effective solution for the malaria problem. Malcolm Watson, who had introduced the concept of 'species sanitation', whereby measures were adapted to target a specific mosquito species, and who had conducted significant anti-malaria work in both Malaya and India, doubted that any amount of education or general sanitation could solve the malaria problem. King viewed the focus on education as a convenient excuse, remarking that the 'true sanitarian in the past in India has been baulked of success largely by the official mind, with its desire for equanimity; with its obsession that "education" of a comparatively small class is the only road to advancement of any sort; with its desire to keep the balance between knowledge and superstition, and between opposed religious susceptibilities'.[60] Watson objected to the 1910 Malaria Commission's conclusion that malaria eradication required active popular cooperation and general improvement of insanitary conditions.[61] He quipped: 'I often think that most of my life has been spent, not in fighting the mosquito, but in fighting the men who were preventing me from fighting the mosquito'.[62] Simply put, to eradicate malaria in India, the government needed to destroy mosquitoes and eliminate their breeding grounds. No amount of education could alter that fact.

The Government of India did not take these critiques or the accusation that quininists dictated policy lightly. It assured the Secretary of State for India that it maintained an open mind on malaria prevention. C. Pardey Lukis, Surgeon General and Acting Sanitary Commissioner with the Government of India, worried about this fracturing of official opinion into two camps. He reassured attendees at the Bombay Malaria Conference in November 1911 that he favored quinine prophylaxis and treatment in conjunction with general sanitation and the elimination of breeding grounds where possible and at a reasonable expense.[63] Lukis held that although quinine had been useful at the individual level, it had conferred little benefit upon the community as a whole. He viewed the problem of malaria as one of mosquito eradication, believing that it was the government's duty to eliminate malaria by eliminating the mosquito.[64] Bentley agreed (despite Ross' labeling him a quininist) and argued that malaria prevention and the complete eradication of malaria infestation could be more quickly and surely realized through indirect methods than by measures targeted specifically at malaria. He saw the improvement of agriculture as the most effective of all anti-malaria measures. F. P. Mackie, Officiating Public Health Commissioner for the Government of India, concurred with Bentley in thinking that India's salvation from malaria lay in a gradual process of bonification, or general agricultural improvements, which included land reclamation by drainage, cultivation, and resettlement.[65] Proper agricultural methods could eliminate stagnant water, and thus mosquito breeding grounds, through proper irrigation and drainage. Bonification would thereby reduce people's exposure to malaria.[66]

Criticism of the government's policies came from outside of the public health establishment as well. The Indian press drew attention to quinine shortages, the high prices of quinine, and black market trading of the drug.[67] Sailaj Lal Chatterjee, head of the Nimta Anti-Malaria Society argued that quinine-centered policies and mass quininization were impractical and had yet to succeed in Bengal. He believed, unlike some officials, that to save Bengal from the 'jaws of malaria', the government needed to educate people on how to prevent malaria themselves.[68] Education about malaria, however, was inadequate and sporadic and failed to reach many of the millions suffering from malaria. An anonymous Indian critic accused both the government and Indian leaders of inaction. He regretted that government worried more about external issues, such as defense, foreign capital, and Indianization of the civil services, and Swarajists and Non-cooperators about independence and debating in Councils than about the malaria scourge.[69]

Indian members of governing bodies also turned a critical eye toward the government's malaria policies. In 1916, Surendranath Banerjee, a leading moderate nationalist and a member of the Bengal Legislative Council, introduced a resolution to the Imperial Council proposing that the Council recommend to the Governor-General that he instruct the provincial governments to take thorough measures to prevent malaria and to publish an annual report on progress made in combatting malaria.[70] Banerjee was persuasive in his choice of arguments; he compared the eradication work of England, Italy, the United States, and Japan to India's and asked, '[Am] I to understand that our Government will confess to a failure in a work in which other Governments have been more successful'? Malaria, he argued, could be prevented if only the government would stop convening more conferences and committees and focus on actively reducing malaria incidence.[71]

Banerjee raised another very important and problematic point: the central government was not holding the provinces accountable for malaria control. The Government of India issued recommendations and formulated policy direction but failed to establish a means to oversee the implementation of policy by provincial governments. Prior to the constitutional changes of 1919, the Government of India still had jurisdiction over public health throughout all of India; yet, it did not ensure that provinces complied with its recommendations to promote quinine use. By contrast, with respect to plague control, the Government of India had been quite diligent in policing provinces to make sure that they abided by the Government of India's policies and the regulations of international sanitary conventions. These divergent approaches toward the provinces reflected the different health, political, and economic exigencies of epidemic plague versus those of malaria.

## Obstacles to policy development

The implementation of malaria control measures in India had been stymied on a strategic level by the Government of India's disease control priorities, which then translated on a practical level into the privileging of further investigation and quinine use over the implementation of sustainable and comprehensive measures. In their defense, officials contended that the 'inquiry and the prevention of malaria are two portions of the same process, and it is a matter for discretion in each case as to where we are to leave the process of inquiry and proceed to that of actual measures affecting the people'. The Malaria Survey of India, a central agency overseeing malaria research and prevention, recommended that any requests for local governments to take action should also include, where necessary, inquiries into malaria prevention. Since the conditions leading to malaria varied across India, each province needed to decide which measures best suited its particular conditions. This position also had the effect of making malaria policy contingent on local authorities and resources.[72]

India's malaria particularities provided a convenient pretext for relegating responsibility for implementing comprehensive anti-malaria campaigns to provincial and local authorities. The Government of India was not about to back itself into a position of responsibility for controlling malaria in all of India. Instead, it continued to emphasize the distribution and use of quinine, since this policy could be applied to all provinces. It also gave grants to provincial governments to support both local inquiries into conditions leading to malaria and the practical application of research toward improving or eliminating those conditions.[73]

In addition to the Government of India's reluctance to engage more fully with the provinces in developing an anti-malaria strategy, the lack of consensus among health experts on how to approach the malaria problem hindered consistent action against malaria. The epidemiological conditions contributing to malaria in India proved so diverse that medical opinion remained divided on which interventions to undertake. Internal disagreements over malaria strategies proved problematic for policy at the central level just as they had for cholera and plague policies. At least with plague and cholera, the Government of India could concentrate on specific sites of control, such as ports, ships, and religious festivals. No such sanitary enclave existed for malaria.

Complicating matters, there were larger social, economic, and political issues that prevented the public health establishment from addressing malaria in more than a superficial way. Indians themselves, according to some public health officials, provided a significant impediment to anti-malaria work. Officials faulted the masses for the lack of progress in improving general sanitation and implementing public health measures. A former Sanitary Commissioner remarked in 1913 that: 'On the one hand Government is committed to a forward policy in sanitation and the educated classes [press] for still further advances and a quicker progress, while on the other, is the great bulk of the people, illiterate, full of prejudice, suspicious of all change and ignorant, as formerly, of even the most elementary rules of hygiene'.[74] King, in his report, 'The Prevention of Malaria in India', wrote that the 'difference of race and finance available locally in various parts of India, the dread of increasing permanent changes, the fear of even indirectly countenancing measures that might be, as conducted by enthusiasts, unwelcome in the face of caste and race prejudices, have combined to bring about the standard sanitary policy of the Government of India – *festina lente'*.[75] Colonial ideas about racial difference manifested in two ways in the formulation of infectious disease policy. Officials were reluctant to implement measures that might violate Indians' views of social custom and status based on ethnic or caste differences. At the same time, health officials used race as an excuse for inaction, arguing that certain habits and customs were ingrained in Indians and could not easily be changed. These attitudes regarding Indians prevented an objective assessment of the feasibility of various proposals. To make matters worse, experts in global health circles validated these negative assessments of Indians.[76] Paul Russell of the Rockefeller Foundation's International Health Division observed on his extensive tour of India that, 'religion frequently bars the way of health officers. They are forbidden to touch this or that malariogenic pool or well, their therapy is scorned, they are reproached for destroying mosquito life, and all too often they are frustrated by dogma and superstition'.[77]

Not all officials held such a contemptuous view of Indians. J. W. D. Megaw, Director-General of the IMS, critiqued these attitudes: the 'prejudice against tackling the problem is chiefly on the part of Europeans who adopt the unreasonable attitude of planning their own existence on rational lines while at the same time they assume that Indians are unable or unwilling to contemplate any change in their outlook on life'.[78] Unfortunately, this particular viewpoint did not prevail in policy discussions.

Health officials became even more cynical after the institution of dyarchy in 1919, which passed responsibility for certain subjects, including sanitation, to provincial and local Indian ministers while keeping subjects such as defense in the hands of British executive councilors. Before the Far Eastern Association of Tropical Medicine, J. D. Graham, India's Public Health Commissioner, remarked on the difficulties of disease control in India:

> Municipal Government which, in these days of the reforms and local autonomy, is largely Indian and very often non-co-operative in tendency, the problem of prevention of disease on modern lines becomes at once surrounded by great difficulties. If we add to these the small numbers of qualified doctors, the ignorance of the population, their prejudices and superstitions, religious and otherwise, the want of a public health conscience in the community and the absence of driving force, it will be realized how difficult it is to promise or even to anticipate.[79]

The Government of India's Sanitary Commissioner painted a picture of public health in India, in which Indians were to be blamed for the abysmal state of health and disease, thus exonerating the government from responsibility for past, present, and future failed public health measures, while also justifying limited government attempts to ameliorate sanitary conditions. Officials displaced the Government of India's failure to address malaria effectively onto local bodies governed by Indians.

The devolution of public health responsibilities to provincial governments also placed the burden and blame for the state of public works onto local Indian representatives. Devolution, with its hollow promise of moving toward 'responsible government' by Indians, was offered to Indian nationalists as a means of assuaging them and the Indian public.[80] The extension of the franchise and decentralization was portrayed as a move to make provincial governments more accountable to the people. In reality, it shifted the burden of additional taxation to Indian representatives and lowered the central government's administrative costs.[81] If local bodies wanted to improve public works, they would need to increase taxation, thereby making it seem that Indians were taxing themselves. The British answer to Indian demands for democracy was to diffuse opposition, communalize and provincialize Indian politics, and protect elite authority by devolving administrative responsibilities to the local and regional level, while keeping real power in the bureaucratic center.[82] It was a guise to give the impression that collaboration could triumph over resistance.[83] Since Indians had lobbied for greater power in local government, officials argued that there was nothing they could do if municipal governments then decided against government intervention in sanitation and disease control. Devolution was a clever ruse to increase British power, not weaken it. Whitehall and the central government needed to retain enough control to ensure that India maintained its creditworthiness and met obligations to remit payments to the metropolis.[84]

While the Government of India's attitudes could and did fetter the implementation of policies, a lack of education to accompany the policies it did institute reduced their impact. Efforts to encourage the use of quinine for treatment and prevention encountered several obstacles. The Government of India and provincial governments did distribute pamphlets about mosquito control in homes and hold lectures and demonstrations, but they did not implement a widespread education campaign to convey the utility of quinine as cure. Education about malaria remained a local effort.[85] As a result, the vast majority of Indians received little if any education about the drug and its benefits. Many did not understand or know how the life cycle of the *Plasmodium* parasite related to fever episodes, and, thus, preferred not to take quinine when not visibly suffering from malaria.[86] This further undercut the government's quinine approach to malaria control.

The Government of India was also reluctant to advertise too widely the benefits of quinine. F. H. G. Hutchinson, Sanitary Commissioner with the Government of India, did not think it prudent to promote education about quinine until production was adequate. The Government of India could not sell quinine at the market rate and widely distribute free quinine until India could manufacture enough quinine and other cinchona alkaloids to meet its needs:

> We are waiting to push quinine, until it is reasonably certain that sufficient quinine will be available to meet the demand result-ing from that push, and that there will be no wide fluctuations in the price of quinine . . . . policy adopted in Italy resulted in a ten fold consumption of quinine in ten years. How is a corresponding response to a campaign in India to be met? It would absorb the whole quinine supply of the world, and unless India can produce the major portion, the market price of quinine would probably jump to an exorbitant figure.

The government applied supply-side economics to quinine use. Until an adequate supply could be secured, demand should not increase. The catch was that until consumption of quinine significantly increased, it would have little influence on malaria in India. The treatment of malaria was as much a problem of market economics as of public health.[87]

Fundamentally, the Government of India did not believe large-scale malaria control economically feasible. Industry, however, had already proven the possibility of combating problems where laborers were concerned. As had been the case with cholera and plague, the Indian Railway Board, the Indian Tea Association, the Indian Mining Association, the cotton industry in Bombay, and other industrial organizations had long understood the importance of disease to their business interests. Malaria mortality and morbidity increased the need to recruit labor from the United Provinces, Bihar, the Central Provinces, and Madras, thus raising the costs to the tea industry of recruiting and training new labor while housing the sick. Tea estates in northeastern India applied malaria control measures along the lines advocated by Ross, while the Mines Board of Health in Jharia and the Asansol Mines Board initiated anti-larval work around mining sites to reduce malaria's impact on coalminers.[88] Just as they had promoted anti-cholera inocula-tions, these industry groups hired malariologists to control malaria in labor camps and during construction and succeeded in diminishing the incidence of malaria among their laborers.[89] Industry had demonstrated that small-scale malaria operations worked. Anti-malarial measures had proven successful on tea plantations, along branches of the state-financed Bengal-Nagpur railway (which was built by the Bengal-Nagpur Railway Company), and in certain ports.[90] Rail was the largest British capital investment in India, and railways were keen to ensure that labor was healthy enough to continue working. The Eastern Bengal Railway Company and the Bengal-Nagpur Railways had both spent large sums annually on larval destruction to protect their workers from malaria along rail lines.[91] In rural areas where industry and commerce were concerned, both chemical and 'naturalistic' controls had proven practicable. F. Senior White of the Bengal-Nagpur Railway had achieved notable success in protecting railway communities in malarious areas through the use of drainage, larvicides and 'herbage cover' methods.[92]

The Government of India thought, however, that the measures used to protect thousands of industrial, rail, and tea workers and their families were too expensive to undertake for the general population. This resulted in central government malaria control remaining primarily limited to the military.[93] While mass malaria control probably was too costly given the colonial government's financial priorities, industry's efforts and controlled experiments among prisoners had at least proven the possibility of reducing malaria mortality.[94]

Provincial governments' targeting of lucrative regions would later demonstrate the possibility of successful government intervention. Provinces had focused on those areas most economically vulnerable to the effects of malaria on labor and production. The Government of Bengal applied larvicides to stop mosquitoes breeding in the industrial rural area near Changail, performed subsoil drainage in the Minglas Tea Estate, and flushed stagnant streams to wipe out larvae in the Topsi-Singaram collieries.[95] The Government of Madras systematically investigated malaria in areas necessary to safeguard the interests of mining and tea syndicates or to protect critical construction projects. During the construction of the then biggest dam in the world on the River Cauvery in the 1920s, the Madras Director of Public Health launched an anti-malarial campaign in the area of construction to prevent any delays due to laborers becoming ill from malaria. This campaign kept the dam project going without delays due to malarial illness.[96] The protection of commercial interests had galvanized provincial health authorities to address malaria. In contrast, the Government of India could not be so easily moved: no direct connection existed between the protection of overseas trade and malaria.

From the 1920s onward, the most significant barrier to effective malaria policy would have nothing to do with epidemiology, access to quinine treatment, or availability of personnel. Rather, the Government of India, by transferring certain state responsibilities to the provincial level with the 1919 Government of India Act, effectively precluded central coordination of and involvement in active malaria control. The act, which went into effect in 1922, formalized and expanded the quasi-federalist structure in India. It affected health policy by transferring jurisdiction in most matters of public health and sanitation to provincial governments.[97] Provincial subjects expanded to include: public health and sanitation; medical administration of hospitals, dispensaries, and asylums; provision of medical education; vital statistics; monitoring of pilgrimage within British India; registration of births and deaths; adulteration of foodstuffs; regulation of import and export trade; and supervision of minor seaports. These responsibilities were subject to infectious disease legislation passed by the Indian legislature, but the central government pushed off onto the provinces most of its public health responsibilities. It retained jurisdiction over: overseas pilgrimages; port quarantine; marine and quarantine hospitals; sanitation for major ports; censuses; and public health statistics.[98] In short, the Government of India maintained the right to oversee the control of epidemics and extra-provincial, interprovincial, and international public health matters.

The division of public health responsibilities between the center and the provinces made it almost impossible for the Government of India to push its quinine policies. Under the new act, the Government of India was constitutionally unable to coordinate and streamline quinine production, pricing, and distribution among and within provinces. Prior to 1919, the Government of India had contemplated bringing all quinine production under its control, but due to the impending reforms of government administration, the Government of India thought this would be beyond its purview and left it to the provincial governments of Madras and Bengal. Realizing that this was problematic, the Government of India worked to amend the Devolution Rules in 1926 to enable it to coordinate cinchona planting and all quinine production in India and to regulate the price, supply, sale, and distribution to provinces of quinine and cinchona. Still, the Government of India could do little to improve distribution and use within provinces, since it had transferred the responsibility of medical relief to the provinces.[99] Provincial governments ultimately had the responsibility of purchasing and distributing quinine, but provincial governments did not have the funds to buy enough treatment, and under the finance rules of the 1919 Act, the Government of India could neither compel provinces to buy quinine, nor favor, and thus aid, any one province by selling quinine to it below market rates.[100]

The Government of India could not eliminate the financial obstacles preventing provincial use of its quinine stocks.[101] Exasperated health officials were 'sick of attempts to get financial co-operation from the provinces in the matter of using stocks acquired for them, would gladly wash [their] hands of all responsibility in these matters and leave the future of quinine to the mercies of a foreign monopoly not always with friendly intentions and that branch of Public Health that hangs on quinine to take care of itself'.[102] So, as the Indian medical press noted, the 'Government of India sit on their quinine, the local governments spend on other measures the money they had put aside for the purchase of quinine, and the *ryot* continues to shake with ague'.[103] Malaria experts had long realized that the lack of coordination among government departments and public health authorities presented one of the major difficulties in malaria control. Attempts to control malaria through quinine, let alone through other measures, did not have sufficient multilevel government support.[104]

The problems coordinating malaria control between the center and provinces were exacerbated by the absence of a central public health agency. Public health policy in India generally lacked uniformity and oversight at the center and resources adequate to the task. F. P. Mackie, the Sanitary Commissioner for the Government of India, lamented:

Under the present constitution the central Government has no power to bring a recalcitrant province to reason nor to insist on its taking even elementary precautions in safeguarding public health. This is an anomaly and is likely to bring grave results in its train. . . . At present the public health activities of a particular province concern that province alone and the legal enactments which exist in one district may be different from those in the next across the interprovincial border. Just as the League of Nations has answered the question 'Am I my brother's keeper'? in the affirmative, where the nations of the world are concerned so it must be realised that in the sub-continent of India provincial boundaries are artificial and are not recognised by epidemic diseases. . . . As it is now, the Public Health Commissioner in Delhi or Simla watches the progress of epidemics passing from province to province without having any official jurisdiction over, or even any official advisory capacity in respect of them.[105]

The League's views on health policy held sway with policymakers in India and influenced their conceptualization of public health infrastructure. Mackie, along with the provincial sanitary commissioners, realized that without a central organization to coordinate provincial public health departments and promote internal cooperation, it would be impossible to check or control outbreaks of infectious diseases within India.[106]

In 1927 the Government of India finally took one step toward establishing central oversight of malaria control by establishing the Malaria Survey of India, a permanent organization dedicated to malaria investigation and prevention, which also served in an advisory capacity to the Government of India on all malaria-related issues. The Malaria Survey assisted provincial organizations in conducting malaria inquiries and could temporarily loan its staff to local governments, connecting malaria work at a central level with that in the provinces. The Malaria Survey advised local officials on the design of centrally-funded rural malaria schemes and maintained close contact with them as schemes were implemented. This reinforced the Survey's aim to promote both investigations and practical applications of research.[107]

In practice, cooperation between the Survey and provincial governments was frequently hampered by the administrative difficulties of coordinating the Government of India, municipalities, port trusts, and railways.[108] The number of authorities involved in malaria control at several levels of administration underscored the need for more comprehensive public health jurisdiction at the center, but with Devolution, which was later reinforced by the 1935 Government of India Act, this was not an option.[109]

Fortunately, Devolution was not wholly damaging to the advancement of public health. It boosted malaria research in India by enabling the Government of India to focus energy and money on malaria studies (instead of on anti-malaria measures) and on building research institutions of international renown. In the realm of malaria research, India had, by the late 1920s, become the darling of the League of Nations Health Organisation.[110] The Government of India and its health officials had made concerted efforts to bring India's research to the attention of the League and to persuade it to expand the League's malaria work to India.[111] Conversely, the League's Malaria Commission hoped that India's Malaria Survey would become an information center, believing its work of great value to all malariologists and sanitary administrations around the world.[112] The League's Health Organisation relied heavily on India's medical experts, requesting that the Government of India send specific malariologists to meetings of the League's Malaria Study Commission to review the results of international inquiries. Ludwik Rajchman, Director of the League's Health Organisation, thought it essential to have a malaria expert from India in attendance, given the importance of the disease in India and the extent and depth of research that had been carried out there during the preceding 60 years.[113] In 1930, Samuel Rickard Christophers, Sydney Price James, R. Senior White,

John A. Sinton, Gordon Covell, and Charles Bentley, all former or then current public health officials in India, accepted nominations to serve on the League's Malaria Commission, highlighting the international importance of India's expertise in understanding malaria.[114]

Was research, however, also an excuse for not addressing the underlying problems of disease? Indian critics of government policies wrote in the *Indian Social Reformer* that they hoped that the League of Nations committee undertaking a tour of India to study malaria would

> not be precluded from inquiring to what extent the poverty of the people and the consequent insufficient nourishment available to them is responsible for the prevalence of malaria by lowering their power of resistance to disease. This, in our opinion, is the crux of the matter, but the Government of India, while prepared to finance expensive research and other schemes which do not go to the root of the matter, is persistently reluctant to face the real issue. That is why Indian public opinion is, as a rule, extremely skeptical of the results of these inquiries which it regards as spectacular displays intended to impress the outer world rather than to improve the health and vigour of the population.[115]

This critique, echoing Banerjee's censure of 13 years earlier, faulted the Government of India for allocating too many resources to malaria research and not enough to malaria control. The League's mandate, however, was not to interfere in national affairs but rather to investigate and report on malaria research and control.[116]

## Rural strategy

The Government of India's research institutions and malariologists may have impressed the League of Nations Health Organisation, if not Indian critics, but its measures to combat rural malaria had not astounded anyone. After concluding its 1929 study tour of India, the League's Malaria Commission remarked that they had seen 'admirable research on malaria control, but we cannot help feeling that malaria control in India should be very much more actively prosecuted as a general duty by the public health departments, and that the control of rural malaria should be taken up more seriously'.[117] They were not alone in their conclusions. V. Shiva Ram and Brij Mohan Sharma, Secretary and Assistant Secretary, respectively, for the Lucknow branch of the League of Nations Union, hoped that the Commission's report would encourage tropical countries, including India, to eliminate malaria and save the lives of its unfortunate victims.[118]

The elimination of malaria among the rural population, about 90 percent of the total population, remained the greatest disease problem facing India. Policymakers faced two fundamental questions in thinking about rural malaria: To what extent was it possible to treat the rural sick? From a strategic perspective, was the objective to be amelioration of the disease's effects or permanent cure?[119] The Government of India and provincial governments could not address rural malaria in any significant fashion without better multilevel coordination and more funding to the provinces for measures to control malaria. Part of the problem lay in the unofficial classification of anti-malaria works into municipal, and thus possible, and rural, and thus impossible.[120] J. A. Sinton, Director of the Malaria Survey of India, stated, in reference to a proposal for a mass treatment experiment in an area of rural Bengal, that given the present state of knowledge, the '"widespread application of anti-mosquito measures for the control of malaria in rural areas in India is not an economically practical proposition"'. Indian critics argued that government malariologists focused on the science of interventions but neglected to ensure that the cost of proposed schemes lay within the budget of the locality.[121] Mosquito eradication was not a viable option either since neither the inhabitants nor local authorities were in a position to finance it.[122] These circumstances resulted in failed schemes and malariologists viewing malaria control in rural areas with despair. They advised villagers to wait until urbanization, when they would be able to afford a public health organization to fight disease.[123] Unfortunately, malaria was not inclined to wait.

In the short-term, the lack of funds left only measures of amelioration and containment. The most immediate solution centered on quinine treatment, since larval control by oil, Paris green (an insecticide), and drainage seemed impracticable both logistically and financially. The Malaria Survey concluded that the best plan was to offer treatment free or at an affordable rate and to make access to treatment within reach of all those afflicted. Since quinine, although not cheap, was the least costly and simplest, if incomplete, answer to the rural malaria problem, both the Government of India and provincial governments chose to improve quinine distribution and popularize its use through propaganda and education.[124] A treatment-centered policy aimed to lower the risk of death, diminish the intensity and duration of attacks, and decrease the resultant period of physical disability.

The Malaria Survey's recommendation for rural India was fundamentally flawed however. Indians living in rural areas did not receive adequate access to the drug. The majority of rural inhabitants could not afford treatment or could not obtain it without difficulty. The cost of quinine treatment to the public was still too high for widespread consumption. Prices for quinine were set by market rates regardless of the cost of production, which was a particularly questionable practice, given that the cinchona factories run by the provinces were reaping enormous gains on the sale of quinine to suffering Indians.[125] Although participants at the 1937 League of Nations Intergovernmental Conference of Far-Eastern Countries on Rural Hygiene had urged health administrators to budget money and energy for malaria control in proportion to the paramount importance of malaria in rural areas, the amount of government money available to treat the millions of rural cases of malaria occurring annually remained limited.[126] Given the high probability of re-infection, and since, as the Government of India believed, most rural Indians preferred not to continue treatment, or could not afford to do so, after the fever and clinical symptoms had ceased, a strategy of permanent cure based on quinine seemed to the government futile and a waste of money and resources.

Quinine distribution to the rural poor, like many public health interventions, had both economic and humanitarian aspects, and the Government of India struggled to reconcile these two dimensions of its treatment policy. The Central Advisory Board for the Government of India stated that: 'From the humanitarian point of view it is desirable that every sufferer from malaria, even the poorest, should be able to obtain sufficient quinine for his needs'. From an economic standpoint, the challenge lay in supplying sufficient quinine at low enough prices to meet the needs of the poor while simultaneously maintaining quinine production on a financially sound basis.[127] For the Government of India, being financially sound also meant that it refused to give away or sell at a reduced price any quinine that exceeded the stipulated reserve quantity, even if this meant that the government risked deterioration of its quinine stockpile for use in epidemic outbreaks.[128]

The impact of the Great Depression on the Government of India's financial resources made the quinine situation even more difficult. In order to balance the federal budget, the government increased taxes and decreased overall expenditures, including outlays on public works. Meeting imperial financial commitments strained India's domestic economy, while a decline in foreign trade, deflationary policies, and the collapse of agrarian prices decreased government income revenues. Because of the economic situation, the Government of India devolved more responsibility to provincial governments, and government public investment in the 1930s remained dismal. At the same time that the government's finances suffered, the social and economic condition of rural Indians deteriorated with the widespread credit crisis and collapse in agricultural markets. Economic distress made quinine purchase a luxury, but perhaps even more of a necessity.[129]

Ironically, the market-based economics of quinine production made self-sufficiency of quinine and febrifuge production a compelling proposition, even if the Government of Madras thought that quinine would never be produced in southern India as cheaply as in Java.[130] Self-sufficiency would have required a considerable extension of cinchona cultivation, since 60 percent of the quinine consumed in India was imported. Despite this obstacle, the Royal Commission on Agriculture, supported this ideal. Self-sufficiency was a prerequisite to providing the cheaper quinine necessary for any large-scale government anti-malaria campaign. But there was a major defect in the Government of India's self-sufficiency approach: the Government of India was constitutionally powerless to assume control over provincial plantations.[131] Without a change in the Governement of India Act, the center could not force the provinces to extend cinchona cultivation to meet rural quinine needs.

In the long term, the lack of financial resources available in or to India's thousands of rural villages meant that the solution to rural malaria would most likely come from naturalistic methods, with special attention to improving agricultural yields, which would then improve the overall health of rural villagers.[132] Malariologists in India and internationally thought that biological and naturalist methods of fighting malaria most effectively diminished the potential endemicity in rural areas.[133] Indian doctors advocated mosquito control through larvivorous fish which cost villagers practically nothing. It was cheaper than other methods of malaria control, did not require training or expert knowledge, and was more permanent in effect.[134]

However, in the late 1930s, the Government of India continued to concentrate on the short term and avoided thinking about the long-term ramifications of postponing adequate malaria control.[135] Even by 1939, government officials had not yet reached a consensus on the future of India's anti-malaria drug policy. The Deputy Director of Agriculture (Cinchona) for the Madras Presidency concluded that India possessed sufficient land to support the cultivation of enough *Cinchona ledgeriana* to provide sufficient quinine (600,000 lbs/year) for an active campaign for all of India. He recommended that quinine, not totaquina or another alkaloid mixture, should be the preferred drug of India's anti-malarial policy and stressed to the Government of India the need to raise cinchona cultivation to the level of Java's. In contrast, S. P. James believed that synthetic drugs were the future of malaria treatment. The Government of India stated that it would put the former's recommendation into effect, but the advent of DDT as a means of malaria control during World War II would make the extension of cinchona cultivation irrelevant.[136]

In some ways, this lack of consensus about drug treatment was immaterial to the larger problem of malaria control. The League of Nations Malaria Commission had concluded, based on all prior experience, that eradicating malaria through prophylactic and curative treatment was practically impossible, because of the inadvisability of administering quinine in large doses for prolonged periods, the logistical difficulties of providing all sufferers with the drugs, and the fact that although quinine and the synthetic drugs diminished morbidity, they could not yet suppress the parasites in all carriers.[137] Until public health officials and malaria experts in India reached a consensus and a workable solution, treatment of malaria with quinine would remain at the core of the Government of India's policies. The rural malaria death rate reflected these realities; it averaged about three times higher than the urban rate.[138] As long as the Government of India maintained quinine at the center of its policies, to the detriment of other potentially more effective anti-malaria measures, it would doom malaria control in India to failure.

## Alternatives to quinine policy

Health officials understood that quinine could not continue to be the center of India's anti-malaria strategy given global production capacity. The Sanitary Commissioner with the Government of India estimated India's need for quinine at 360,000 pounds, emphasizing that present consumption of quinine could not be taken to represent actual need. Other estimates placed the amount of quinine necessary to treat malaria in India effectively anywhere from 500,000 to 1,500,000 pounds, but the total world output of quinine was only about 1,000,000 pounds, with nine-tenths of the cinchona bark providing the quinine coming from Java plantations.[139] The use of other cinchona derivatives only had a limited impact as well, given that no more than 10–15 percent of malaria cases received any treatment, either with quinine (an isolated cinchona alkaloid) or cinchona febrifuge (a mixture of cinchona alkaloids).[140] Cinchona febrifuge was still too costly for most, and an improved production to lower prices was years away.[141]

Secondly, existing supplies of quinine and febrifuge did not reach those most in need. The majority of quinine was purchased by well-off persons and unscrupulous traders. Unregulated free traders sometimes adulterated quinine, while post office pice packets remained expensive and in competition with agencies distributing quinine at no charge. The Government of India did not think it could reduce prices because of the costs of planting, manufacturing, and distribution, and its prices between Rs. 12–18/lb were already well below the world market price of Rs. 22/lb. Price reductions in 1928 and 1932 had disappointingly failed to increase consumption, and the cost remained too high to induce provincial governments to buy more quinine.[142] In addition, provincial governments could not afford to sell the quinine they already had below cost.[143]

Thirdly, debates about the efficacy of various cinchona alkaloids and synthetic drugs further complicated quinine policy. Febrifuge was cheaper to produce than quinine, but the lack of scientific consensus on the therapeutic efficacy of cinchona febrifuge prevented extended cultivation in India of *Cinchona succirubra* trees, the species yielding the greatest percentage of total alkaloids. Until medical opinion 'reached stability', as to which species of tree would prove most robust in terms of alkaloids and viability, the government could not formulate a new cinchona planting policy in the hopes of reducing the market price of quinine and other cinchona alkaloids. Increasing the supply and use of cinchona febrifuge was crucial however, given that the Dutch-based Quinine Manufacturers' Association sought to control the price of quinine on the world market through calculated limitations on output or by withholding quinine stock whenever production exceeded consumption.[144] Synthetic drugs had drawbacks as well. Atebrin and plasmoquine, both developed in the late 1920s, were expensive and had uncomfortable and toxic side effects, including yellow coloration of the skin, and, thus, were unlikely to displace quinine and other cinchona alkaloids from general use.

The establishment of the efficacy of febrifuge in the late 1920s still did not resolve the quinine versus febrifuge debate. Research by William Fletcher in the Federated Malay States and by the League of Nations had shown that non-quinine cinchona alkaloids treated malaria as effectively as quinine. All that was needed in India was a change of medical opinion in favor of these other alkaloids to encourage the cultivation of cinchona trees, thereby allowing production of a treatment at a cost attainable to the poor.[145] The Government of India was slow to react, arguing that the composition required standardization before it could be put into use, but even after the League of Nations malaria committee had devised a standard composition for febrifuge, known as totaquina, government factories did not begin manufacture and had yet to conduct clinical trials in India.[146] Bureaucratic inertia combined with skewed priorities and a central government out-of-sync with provincial needs to produce more of the *festina lente* King had so criticized.

Using a peculiar method of reasoning policymakers concluded that since the public demanded quinine and not febrifuge, the government should shape policy to demand and not vice versa. This rationale reflected economic interest and lack of political will rather than sound medical opinion. Alexander J. H. Russell, Public Health Commissioner with the Government of India, justified a quinine-preferred policy by arguing that if cinchona febrifuge became popular, then its price could rise to that of quinine, thereby eliminating the benefits of its manufacture. He was duly criticized by the medical press in India, which accused the medical profession of perpetuating the popular preference for quinine over febrifuge.[147]

The government was also responsible for this preference, for reasons which extended beyond questions of therapeutic efficacy to include less palatable ones of government finance. The Government of India wanted to ensure a market for its quinine stocks. First, it allotted to provincial governments a quantity of febrifuge in proportion to the amount of quinine they purchased from the Government of India, since febrifuge was a by-product of quinine manufacture. Second, the Government of India prohibited provincial governments from purchasing quinine or febrifuge from abroad, once again pitting the provinces against the center. In the mid-1920s, when Bihar and Orissa and other provinces faced difficulties providing and extending anti-malaria treatment due to a febrifuge shortage, the Government of India asked provincial governments to buy the more expensive quinine to meet their needs. In post-Devolution India, the government maintained that in cases of epidemics, it was the responsibility of the provinces to increase grants to state and state-supported institutions to enable them to purchase more quinine. It was not for some provinces to break away from an all-India arrangement on the purchase of quinine and febrifuge the 'moment they found another course less burdensome to their finances'. The Government of India would not allow Bihar and Orissa to purchase febrifuge from the Java Combine because it would have violated the Government of India's policy of gradually making India independent of foreign febrifuge or quinine, and it would have eliminated Bihar and Orissa as buyers for government quinine.[148] Thus, the Government of India constrained the Government of Bihar and Orissa to purchase more costly quinine from Indian stocks.[149]

Official medical opinion in India regarding quinine did begin to change in the late 1920s and 1930s as policymakers increasingly acknowledged that quinine could ever only be one aspect of malaria policy. Public health experts realized that quinine treatment, while important as a therapeutic, could not prevent further infections, and at best was a temporary measure. Mackie argued that although curative medicine was necessary on philanthropic and humanitarian grounds, it would only ever be a secondary factor in determining national health. The key to disease eradication lay in prevention.[150] While quinine could greatly reduce the severity and incidence of the disease, malaria experts agreed that quinine could not always prevent infection, regardless of how routinely it was taken. Its chief effect was to lessen the severity, duration, and fatality of malarial attacks.[151] Anti-mosquito measures, in contrast, could be permanent or at least of long duration. Furthermore, officials thought quinine treatment would prove more costly in the long term than anti-mosquito measures.

While policy continued to focus on quinine, officials realized that proper sanitation, the elimination of mosquito breeding grounds, the provision of piped water, and adequate nutrition were just as, if not more, critical to fighting malaria. Nutritional research had suggested to health officials that the 'poor physique' and lack of resistance to disease of a significant portion of Indians resulted from a 'badly regulated diet prescribed by custom' and had highlighted the importance of a well-balanced diet in resisting endemic disease.[152] Hehir argued that the prevention of malaria was intertwined with and inseparable from improvement in the economic conditions of rural Indians.[153] Surendranath Banerjee also viewed poverty as largely responsible for the prevalence of malaria in Bengal. It weakened the body and reduced its power of resistance against disease. He drew a positive correlation between agricultural prosperity and the health of different parts of Bengal, and argued that the government must prioritize the reduction of poverty if it wished to achieve a reduction of malaria.[154]

Malariologists believed that the economic and social conditions of the masses, the quality of people's diets, and the incidence of malaria were interrelated. Experts had demonstrated a relationship between the average prices of grains and the eruption of malaria epidemics. Low food availability or high food prices led to conditions of famine and stress, which then increased susceptibility to malaria and other diseases.[155] Hehir believed that dealing with malaria effectively was not simply a matter of eliminating stagnant water or distributing quinine.[156] Research on the connection between nutritional status and malaria incidence made the government's focus on quinine appear even more shortsighted. It also lent more weight to a comprehensive approach to malaria that could address both the human and vector components of malaria transmission. The Public Health Commissioner with the Government of India believed that no malaria prevention campaign was likely to be successful unless those responsible realized the vital importance of defective nutrition in increasing susceptibility to malarial infection.[157] But if mosquito eradication and quinine distribution had their logistical and financial difficulties, addressing the nutritional state of Indians living in malarious areas was a herculean task. It was one that the Government of India was neither willing nor able to confront with existing resources, especially since officials had construed malnutrition as an endemic problem for which the colonial state had little responsibility and over which it could exert little control.[158]

Even if the Government of India did not address the nutritional status of India's population, experience had shown that malaria incidence could be reduced by employing other strategies. Defensive measures could decrease human susceptibility and offensive measures could eliminate parasites and their vectors.[159] Evidence supporting these beliefs resulted in the advocacy of a more multipronged approach to malaria control. The use of oil on stagnant water, Paris green, and quinine were increasingly seen as stopgaps, necessary but insufficient in the long term. The studies of Bentley, Strickland, Christophers, and Sinton demonstrated that the future of anti-malaria campaigns lay in biological methods, including drainage, irrigation, the use of larvae-eating fish, and planned releases of dammed water.[160]

Practically, however, official policy continued to emphasize quinine treatment over anti-vector and anti-parasite controls, even though the Government of India had begun experimenting with the use of larvivorous fish as early as 1900.[161] Almost all of the anti-malaria measures carried out in the provinces for the prevention and destruction of mosquitoes were temporary, and did nothing to decrease reliance on quinine.[162] These policies were backed by the League's Malaria Commission, which concluded in 1927 that in endemic areas, where the acquisition of partial immunity conferred health advantages, countries should endeavor to reduce the severity and fatality of malaria through quinine use rather than undertake measures, which would be necessary were the goal to eliminate parasites completely from the affected area. The Commission, on which S. P. James played an active role and

which, in 1930, consisted of six prominent Anglo-Indian health officials, viewed malaria as a social disease most effectively fought through
social improvement and the use of quinine. It dismissed long and costly
anti-mosquito campaigns. However, Malcolm Watson, who had since
been shipped off to Malaya where it was hoped he would be less critical
of the Government of India's policies, and the Rockefeller Foundation
disagreed and continued to advocate anti-anopheles campaigns.[163]
Thus, into the late 1930s, despite the fact that 1.25 million Indians were
dying from malaria annually, the Government of India, conveniently
supported by an Anglo-Indian dominated League of Nations Malaria
Commission, carried on with its policy of amelioration rather than
of eradication or comprehensive control and continued to focus on
increasing quinine consumption.[164]

## Conclusion

Between the late 1890s, when Ross elucidated the relationship between
the *Plasmodium* parasite, the mosquito, and the transmission of malaria,
and the 1940s, when DDT came into use as an insecticide, India had
paradoxically progressed to becoming an important authority internationally on malaria, while failing to achieve any marked improvement
in the reduction of malaria morbidity and mortality. The elaboration of
the mode of malaria transmission during the 1880s and 1890s had created a new framework for viewing the disease – one that narrowed the
scope of preventive and treatment measures to attacking the parasite
in man and reducing the mosquito vector to the detriment of addressing the social and economic factors affecting malaria incidence.[165] This
biomedical approach began to shift in the 1920s and 1930s toward an
emphasis on improving the social and economic conditions of Indians,
but by that point, the Government of India had largely devolved public
health responsibilities to provincial and local governments. Ironically,
the very legislation which aimed to move India toward 'responsible government' had hindered health progress domestically by leaving central
health officials powerless to force provincial governments to adopt certain malaria control measures. The unintended consequence was that,
with the bulk of public health responsibilities being relegated to the
provinces, the Government of India could concentrate on international
public health affairs and research. The international community praised
the malaria research coming out of India and frequently consulted
India's disease experts, even while recognizing that the central and provincial governments could do more to control malaria.

A comparison of the Government of India's approach to malaria versus that to cholera and plague elucidates in part why this paradox existed. On a practical level, malaria, cholera, and plague policy development were all structured by resource availability and officials' attitudes toward Indians. On a high level, differences in policy stemmed from differences in epidemiology and the potential risks arising from a specific disease. The types of risk, the way in which those risks were perceived, and the ramifications of addressing or ignoring those risks and the spread of disease varied by disease, political climates, economic priorities, and the amount of political will to control disease.

Malaria epidemics were not accorded the same importance as cholera or plague outbreaks by the Government of India or by European states. It was epidemiologically unlikely that a malaria epidemic in India would find its way to Europe, and even if it could, southern Europe was already infected and heavily invested in reducing malaria incidence. While malaria laid waste to millions of lives, crippled labor productivity, and impeded economic development, it did not affect the flow of India's international trade and did not cause public disorder.[166] The set of issues that had made cholera and plague epidemics urgent matters to be dealt with systematically did not apply to malaria. Trade, international politics, and global epidemiological security never factored into malaria policy discussions.

During epidemic outbreaks of plague and cholera, the Government of India sought to minimize its exposure to risk – whether commercial, political, or epidemiological. The central government focused both central and provincial funds and energy on containing plague and cholera outbreaks in an effort to appease European states. Malaria, like endemic cholera and plague, evinced a much less decisive reaction, even though it was more critical to address in terms of morbidity and mortality. Thus the Government of India's approach to India's malaria problem was more influenced by political will, Devolution, colonial economic priorities, and the economics of pharmaceutical supply and demand than it was by the commercial, epidemiological, or political risks of failing to address a particular public health issue.

Policies for these three diseases did converge on a particular issue, however. Investigations of malaria, like those for cholera and plague, helped the government achieve its goal of carving out a position of prominence for India in international health circles. The same jockeying for international prestige that had driven the establishment of India's first research institutions had also fueled the government's support of malaria investigations. As a result, starting in the 1890s, research became integral to malaria policy. Patrick Manson, supported by Joseph Chamberlain and the Colonial Office, had argued that tropical medicine, and a cure for malaria in particular, would benefit the empire tremendously, relieve financial burdens, increase efficiency, and reduce the cost of commercial enterprises.[167] This early emphasis on research, though, hurt the agenda of Ross and others, who wanted to promote a more holistic anti-malaria strategy, which included but was not limited to laboratory-based solutions.[168]

After 1919, the devolution of public health and sanitary matters would constrain an already narrow anti-malaria approach. Secretary of State Montagu and Viceroy Chelmsford had, with the 1919 Government of India Act, translated what was lack of political and financial will pre-1919 into lack of jurisdiction post-1919. This transfer of responsibility from the center to the provinces permitted the Government of India to focus on research and quinine as the mainstays of policy. Thus, the government's overarching policy emphasized amelioration, not prevention or long-term cure, despite opinions that at least in urban areas, malaria control could have been a 'paying proposition' or in today's terminology, cost-effective. In towns and cities, malaria caused greater financial loss than the sums needed to eradicate or control the disease.[169] The League's Malaria Commission, after its 1929 tour of India, criticized the Government of India for having failed to address the '"simple clear-cut conditions of urban malaria"', while the Royal Commission of Labour in India found that 'only too often action on health matters ends with the holding of an investigation and the writing of a report, little effort being made subsequently to carry out even the simplest of its recommendations"'.[170] Even the Government of India's sanitary commissioners decried the lack of progress in treating urban malaria.[171]

This is not to say that the Government of India sat by and did nothing except conduct investigations and distribute quinine, even though this may have seemed the case to some of its critics. It did direct and fund anti-malaria schemes in the provinces to eliminate mosquito breeding grounds and attack the parasites in people through quinine treatment.[172] However, rather than undertake large-scale subsoil drainage, mosquito elimination, and mass quininization to eliminate malaria, the colonial government preferred small-scale measures, trials, and experiments.[173] The Government of India's medical officers did perceive a real economic and humanitarian need to address malaria with social, economic, and medical solutions. Medical workers emphasized the monetary and productivity costs of malaria to India and underscored the demoralization accompanying the debilitating effects of endemic, hyperendemic, and epidemic malaria. The disease posed a tremendous internal risk to India's social and economic advancement, yet policy remained focused on the quinine quick fix.

A quinine-centered anti-malaria policy could never be a long-term solution to India's biggest and most difficult disease problem, especially given the Government of India's inability to obtain, produce, or sell enough quinine to treat more than 15 percent of those suffering.[174] The drug did not significantly reduce transmission or long-term incidence, and, as early as 1900, the international scientific community knew that quinine could not kill the sexual forms of *P. falciparum*.[175] The Government of India's quinine strategy was also problematic given the probability of repeat infection if the environmental conditions (mosquito breeding grounds, improper drainage and irrigation, lack of piped water) and predisposing factors (undernutrition, malnutrition) remained unaddressed or only haphazardly so. This policy made even less sense given that Government of India and League of Nations health officials believed that the value of quinine prophylaxis was greatly limited by India's 'free population, the ignorance and apathy of the masses, their prejudice against the drug, their objection to medicine when not actually suffering from illness, and the fact that it must be continued over an indefinite number of years'.[176]

Malariologists continued to insist that malaria control could best be achieved by both increasing the resistance of the human host and decreasing the amount of infection. Medical officers who had dedicated their professional lives to understanding malaria believed that the reduction of endemic malaria was feasible worldwide, but India would not obtain this goal as long as governments continued to 'toy' with malaria measures, and that only perfunctorily.[177] The long-term interventions this approach required demanded much more of the Government of India than it was willing or able to give. A comprehensive malaria strategy required cooperation between central, provincial, and local authorities; greater authority and responsibility in the center; and a multipronged approach utilizing chemical, biological, and pharmaceutical methods.

The complexity of malaria's epidemiology and its ties to social and environmental factors meant that malaria control also required structural remedies, including the improvement of construction works, agricultural practices, and the social, nutritional, and economic conditions of Indians. Health experts throughout Asia recognized that rural reconstruction, which included improved diets, sanitary dwellings, and education, was necessary to combat malaria.[178] The medical community in India in the 1900s came to believe that large-scale malaria prevention would require changes in cultivation methods and thereby probably a change in the structure of the colonial economy.[179]

Overhauling India's modes of production and the country's health status was beyond the Government of India's financial capabilities and political will however. Funding cinchona agriculture and malaria control measures did not factor into the Government of India's primary fiscal goals of ensuring the viability of India as a market for British goods and capital exports, repaying interest on debt bonds, making remittances for Home Charges, and maintaining the Indian Army. In the late nineteenth century, debt service and Home Charges amounted to about 16 percent of India's revenues; by 1933 they had reached a peak of over 27 percent. The ability of the Government of India to meet these three objectives depended on balancing internal and external demands for its revenue resources.[180] Disease prevention and control simply did not receive enough priority, and malaria did not carry the type of risk that mattered to the Government of India.

Malaria prevention and treatment did not top the government's list of infectious disease priorities before or after Devolution. There were neither political nor trade risks associated with malaria, and the loss to human life had not proved compelling enough. If there was little hope before 1919 that the Government of India would approach malaria control more holistically, there was even less reason for optimism after public health had been officially relegated to the provinces. Devolution conveniently provided justification for the continuance of an already outdated strategy to control malaria. It would also hamper attempts to prepare for yellow fever, as the Government of India found itself for the first time trying to prevent a major infectious disease from invading its borders.

# 4
# From Panama to Khartoum – Yellow Fever Inches Closer to Home

Unlike plague, cholera, and malaria, yellow fever never made it into India.[1] Indians were fortuitously spared the ravages of hemorrhagic fevers and the unsightliness of black vomit and jaundiced bodies. During the first half of the 1900s, government officials could not have known that India would be so lucky. They dreaded having to battle yet another infectious disease and the potential public health disaster and consequent damage to trade that might ensue if yellow fever gained a foothold in the population. Conditions in India could not have been more propitious for the propagation of this disease: a 'non-immune population living in intimate contact with an efficient insect carrier, and an abundance of susceptible animals capable of acting as a reservoir for the virus, together with a climate suitable for the transmission of the disease'. Only strict preventative measures to protect India's borders could prevent what public health officials assumed would 'constitute a calamity of the first importance'.[2] The Government of India recognized that if yellow fever entered the country, an epidemic would be nearly impossible to prevent. The Government of India's anxiety about the possibility of a new epidemic disease invading India, along with concerns about the increasing connectedness of endemic and non-endemic zones, significantly influenced the kinds of policies the government chose to implement. India's approach to preventing the ingress of yellow fever provides insight into how a government's fear of the unknown can be a powerful justification for infectious disease policy.

## The Panama Canal threat

Yellow fever, more than any other disease then known, changed India's relationship to disease control. For the first time, India needed to defend itself and reverse the direction of border control to prevent the introduction of a major communicable disease.[3] The government's task of prevention was complicated by advances in global transportation: first, with the opening of the Panama Canal and later, with increasing airplane travel from West and East Africa. Although yellow fever had been known to exist in the Americas for hundreds of years, the disease had not concerned India until improved and faster methods of maritime and air transport meant that yellow-fever-infected mosquitoes or people could feasibly travel from infected areas to India within the incubation period of the disease. The impending opening of the Panama Canal would significantly increase the interconnectedness of tropical countries around the world, presumably facilitating disease spread. By passing though the canal, ships traveling from endemic areas in the Americas could potentially reach Asia more quickly than if they were to sail east via the Suez Canal.

The opening of the Panama Canal raised concerns about yellow fever spreading to India via China. In 1903 Patrick Manson, one of the earliest and most influential parasitologists, had sounded the first warning about the danger of yellow fever importation into the East from the West Indies and Equatorial America upon completion of the canal.[4] Bombay, the largest port in India, possessed two of the three main factors conducive to the propagation of the virus – an abundance of the mosquito vector, *Stegomyia fasciata*, and a completely non-immune population; only the virus was lacking. Plus Indian laborers in the West Indies had proved susceptible to the disease, thus eliminating the comforting possibility of an ethnic immunity. Doctors and scientists in Bombay suggested that the government send medical officers to eradicate all mosquitoes from major ports in India, thereby removing the vector, and send officers to endemic areas to study clinical aspects of the disease. Some officials, however, thought that defensive measures in India would be too costly and 'irksome' and not necessarily reliable.[5]

The Bombay branch of the British Medical Association realized that preventive measures could not be limited to the Indian peninsula. Participants at the 1910 meeting, with the support of the Sanitary Commissioner of Bombay, called for an international conference to organize an international campaign against *Stegomyia fasciata* in endemic centers, especially near the proposed canal.[6] This suggestion had novel implications – the prevention of yellow fever in India necessitated active disease control outside of India's borders.[7] Whereas India traditionally had held the position of being the source of many epidemic diseases and thus needed to be regulated to protect Europe, now the Americas and West Africa needed to be controlled to protect India. The opening of the Panama Canal placed India on the defensive as health officials found themselves in uncharted public health waters.

The Government of India appreciated that India needed protection, especially since an international campaign to eradicate yellow fever from all endemic areas seemed improbable. Both the Government of India and provincial governments were acutely aware of the disease implications of shortening the time needed to sail from endemic areas to Asia. The Government of India reacted by instituting what British health officials deemed to be antiquated and irrational quarantine regulations. The Local Government Board (LGB) in London reviewed the Government of India's quarantine rules and found 14 days' quarantine unreasonable, since six days was the internationally accepted maximum incubation period of yellow fever in humans. The Government of India refuted the LGB's claim, stating that recent investigations had shown that the maximum incubation for yellow fever was 13 days. Diagnosis of mild cases was extremely difficult, and these people could be infectious for three days. As a further precaution, the Government of India had proposed setting the mooring distance of ships coming from infected areas at a minimum of two miles from land. The LGB emphasized that this would involve serious inconvenience to shipping, could endanger vessels, and might not be practicable for all ports. The LGB pointed out that the Second General International Sanitary Convention of the American Republics of 1905 had stipulated only a 200-meter mooring distance for yellow-fever-infected ships. The Government of India retorted that since the introduction of the disease into India would be disastrous, no avoidable risks were to be taken. As a small concession to the LGB's concerns, the Government of India left it to the discretion of provincial port authorities to fix the distance from shore at which yellow-fever-infected ships should moor, but it recommended a distance of at least one mile. As compensation for decreasing the mooring distance, the Government of India increased the quarantine period for healthy persons to 17 days – a move that did not sit well in London.[8] The Government of India was ready to countenance some inconvenience to trade if it could protect India's ports from yellow fever.

At the same time that the Government of India put into effect these quarantine procedures, it recognized that it needed a better understanding of the disease and of exactly how the Panama Canal would affect maritime transport and, thus, the potential spread of yellow fever. The newly established Indian Research Fund Association decided that its first task should be to inquire into methods of preventing yellow fever from entering Indian ports and of eradicating the disease should it enter.[9] The government also deputed S. P. James to Hong Kong in 1912 to attend a conference on yellow fever. From Hong Kong he traveled to the Americas to observe the disease in its endemic home and to study measures for its prevention and eradication.[10] By visiting endemic areas, James could take advantage of the experience and expertise of health officials and scientists in the Americas to understand how to prevent the disease from taking hold in India. James returned to India, sailing the same route that ships would take to proceed to India once the canal had opened. He reported that, given the longer distances traversed to reach India via the Panama Canal, it was not practical to transact trade with India through the canal. Thus there was no immediate danger of yellow fever being imported into India upon the canal's opening.[11] Rather, the closer prox-imity of the Far East and South Pacific to the canal meant that the danger to India was secondary to and dependent upon the infection of ports in Japan, China, the East Indies, or Australia. The more immediate concern was the spread of yellow fever to these countries and not to India.[12]

Although the situation in the Panama Canal Zone had improved in the interim between 1903 when Manson had first called attention to the yellow fever peril and 1912 when James visited the area, James urged the government to remain vigilant. He warned that although the disease had been eliminated from Panama, Colon, Havana, New Orleans, Rio de Janeiro, and nearly all of the West Indies islands, the Government of India could not rest at ease.[13] It still needed to reduce the numbers of Stegomyia mosquitoes in Indian ports. James advocated the construction of a continuous supply of high-pressure, piped water at all major seaports as an essential first step to eliminate the need for water receptacles, which were the preferred breeding ground of the mosquito vector. James also argued that improving the water supply was more likely to be supported by the public than other measures used in the Americas, such as compulsory inspection and fines for finding larvae on one's premises. A continuous water supply would also help eliminate objections to closing wells, which for both practical and reli-gious reasons, many Indians opposed. In the meantime, while consider-ing James' proposals, the Government of India deputed medical officers and subordinates to carry out surveys of Stegomyia prevalence in India's chief ports (Calcutta, Bombay, Madras, Karachi, and Rangoon) with a view to destroying this mosquito vector in case the first line of defense – inspecting ships coming from infected areas – should break down.[14]

The 'yellow fever peril' was also occupying the minds of interna-tional health authorities. In 1912, the same year as James' visit to the

Americas, they convened an International Sanitary Conference to establish international regulations regarding yellow fever. The convention sub-committee determined that yellow fever was not of great practical interest for European countries because the danger of the disease extending to Europe was minimal. The disease tended to stay within certain latitudes, and if yellow fever did occur on ships, it usually disappeared when the ship reached cooler climes.[15]

The British government's stance at this convention vis-à-vis its colonies would prove critical to India.[16] The British delegate, Lancelot Carnegie, declared that the stipulations of the convention would not apply to any of the colonies, possessions, or protectorates of the British Empire.[17] The British government reserved for each of its colonies and other territories the right to adhere to the convention or to renounce it without being tied to the decisions of the British government. This political autonomy facilitated the Government of India's ability to maneuver within the international public health arena and to institute health regulations as it saw fit. Since yellow fever presented a greater risk to India due to its Stegomyia population than to most of the countries signatory to the convention, India reserved the right to adopt measures more stringent than those outlined in the convention.[18]

The Government of India took full advantage of Britain's liberality and declined ratification. Instead, it retained strict quarantine rules, much to Britain's chagrin. India's quarantine regulations violated accepted practice, and furthermore, threatened the public health diplomacy of Britain and jeopardized the power of British delegates at future conferences. The LGB argued that British delegates had lobbied for more moderate convention measures, and it would be embarrassing for His Majesty's Government to then make reservations on behalf of India due to the convention's leniency. The British government criticized the Government of India's reservations as constituting a return to the ineffective system of quarantine, which Britain had abolished in 1896. It argued that the Government of India's abandonment of the internationally accepted principle of basing regulations on the health of the ship and *not* on the health of the port undercut 25 years of lobbying by British delegates for the former.[19] This action by the Government of India, which would be linked to Britain in the minds of foreign delegates, could potentially reopen discussions of plague and cholera regulations. Britain could not argue one set of principles for other countries, and then when its territories were in danger, adopt more convenient ones. Many countries were also keen to revert to old quarantine practices based on the health of the departure port, and India's behavior left

Britain no grounds to combat reactionary proposals that might hinder British shipping. The British delegates to the sanitary conference, the LGB, the Foreign Office, and the Board of Trade all concurred in condemning the Government of India's exceptionalism.

Nevertheless, the Government of India maintained its position, despite one of its officials having concluded that the opening of the Panama Canal did not considerably increase the risk of yellow fever importation. The Government of India argued that the severe consequences of the disease's gaining entry justified the government's extreme precautions and its refusal to relax regulations.[20] Health officials in India argued that quarantine regulations against yellow fever could not be dispensed with, since sanitary conditions at Indian ports remained unsatisfactory generally and in terms of Stegomyia prevalence specifically. Quarantine would have to compensate for inadequate port sanitation. The government seemed willing to tolerate some inconvenience to trade in instituting precautionary measures, if it meant that it could prevent yellow fever from entering India and causing greater economic damage later.[21]

For the Government of India, the threat of yellow fever overrode both international and imperial public health politics – a position which stood in marked contrast to the influence of politics on early plague and cholera policies. Havelock Charles, President of the Medical Board at the India Office and Medical Adviser to the Secretary of State for India, told the LGB that the Government of India felt warranted in instituting more stringent quarantine measures because of the possibility that Stegomyia could survive on ships for six weeks and thus travel from ports in Central or South America to ports in India. This accounted for the Government of India's rule that any ship touching a yellow-fever-infected area within two months of arriving in India should be classed as suspect, a definition at odds with the internationally accepted one. The LGB argued that the minimum distance from India to a Central or South American port was 11,000 miles, which was long enough for a case of yellow fever to appear if someone was infected. The board also pointed out that yellow fever incidence was waning in endemic areas, and even one of its officials, S. P. James, believed that anti-Stegomyia measures at ports remained the best prophylactic. Charles countered saying that 'peculiar difficulties' due to the religious and political prejudices of Indians had hindered mosquito elimination at ports. The LGB further objected to the Government of India's decision to quarantine all infected persons for the first four days of illness, since keeping ill persons on board contravened the internationally accepted principle that all infected sick should be immediately landed and isolated and treated on shore at a hospital. Charles replied that, because not all Indian ports had hospitals and because anti-Stegomyia campaigns in India were financially impracticable, the Government of India found it necessary to resort to quarantine regulations. Despite the possibility of damaging the authority of the British Government and its delegates, the possible breakup of international agreements on plague, cholera, and yellow fever, and a possible reprisal by certain foreign governments on ships and passengers coming from India, the Government of India would not be swayed.[22]

Although Charles recognized that British delegates would have a tougher time with negotiations at future International Sanitary Conferences, the 'sole object of the Indian Government was to keep yellow fever out of India. No considerations of trade and commerce would be allowed to affect the attitude adopted by the Government of India, though it was shown that India's overseas trade in the fiscal year 1913–14 amounted to no less than £327,225,320'.[23] C. Pardey Lukis, Director General of the IMS, agreed entirely with Charles' opinions and attitudes: 'We can't run the risk of a repetition of our 1896 experience with plague. There was some excuse then as nothing was known about the disease: we should have no excuse at all if we allowed yellow fever to obtain a hold'.[24] Whereas the immediate protection of trade had been the major impetus behind controlling epidemic plague and cholera, it would not be an important factor in the Government of India's decisions regarding yellow fever policy. Instead, an unquantifiable epidemiological risk would shape the government's response to the yellow fever threat.

The Government of India's actions and inactions also came under fire from its own health officials. A sense of urgency pervaded the Government of India's discussions at international conferences and internal communications, yet central and provincial governments were slow to implement definitive domestic precautions against yellow fever's ingress, particularly after public health responsibilities had been devolved to the provinces. F. Norman White, who would later become the League of Nations Health Organisation's chief epidemic commissioner, pointed out to the Government of India the danger of postponing yellow fever precautions until the war was over and the fallibility of thinking that India had a month's grace period from the time the first case appeared in the East until the disease reached India. Two weeks could easily lapse between the arrival of an infected patient and the first indigenous case, leaving India with very little time to prepare. Furthermore, quarantine arrangements were not satisfactory in

Hong Kong, the most likely jumping off point for yellow fever spread to India. The war had also delayed detailed consideration of a yellow-fever notification system between India and other British colonies.[25]

In some ways, the Government of India did not have an acceptable alternative to imposing reactionary quarantine rules, since none of the provincial governments had been able to implement measures to protect their ports against yellow fever. Due to the financial stringency caused by the war, Madras' operations to improve the city's water supply had been considerably restricted, while the Government of Bengal was not in a financial position to fund a quarantine station. The Government of Bombay also found the situation much more difficult to control than had been anticipated and thought that the elimination of Stegomyia from all of Bombay city would be a 'hopelessly impracticable task on account of expense'.[26] Furthermore, the matter seemed less urgent to the provinces due to the temporary closure of the Panama Canal. The Government of India faced a dilemma: if yellow fever was imported, the consequences would be disastrous; yet, the major provincial governments were unable and/or unwilling financially to implement adequate precautions. But since James' report had stated that there was no imminent danger and the canal was temporarily closed, the Government of India and provincial governments thought they could afford a little laxity for the time being.

## Domestic preparedness stalls

With the reopening of the canal in 1917, the Government of India once again thought it imperative for provincial governments to implement measures for both prevention and mitigation in the event of yellow fever infection, but the financial relationships between the center and the provinces undermined the Government of India's ability to implement effective policies, as they had done in the cases of plague, cholera, and malaria. The Government of India wanted the Government of Bengal to establish a suitable quarantine station, yet the Government of India was unable to offer Bengal financial aid from its imperial revenues. The Government of India suggested that Bengal draw funds from its large balances.[27] The Government of Bombay had also asked for help with the heavy expenditures necessary to make effective quarantine arrangements for yellow fever in Bombay, since costs could not be met from port or provincial revenues. The Government of India replied in the negative, stating that the 'mere fact that the service might be one of some Imperial importance does not *per se* constitute a justification for compensation from Imperial funds to the Provincial Governments on which that particular service is thrown. Various services for which provincial revenues are habitually held responsible have a more than provincial importance'.[28] The Government of India was unwilling to provide funding for port preparedness in the provinces, even though vulnerability in just one province constituted a grave risk for all of India.

Tensions between the central and provincial governments over the financing of port preparedness were further complicated by the devolution of public health responsibilities that occurred after the Government of India had initially proposed the establishment of isolation stations in 1916. Before the 1919 Act, it was accepted that provincial governments would bear the financial responsibility of equipping ports. With the passage of reforms, quarantine was clearly defined as a central subject, leaving the Government of India to accept ultimate financial responsibility for port isolation stations. Under the new act, the Government of India was prepared to accept liability for financing an isolation station and observation camp at Calcutta, but it regretted that in 'view of the need for restricting the provision of new civil expenditure to objects of imperative necessity', the Government of India was unable to provide the full amount needed.[29] Despite the change in jurisdiction, the Government of India remained unwilling to fund isolation stations, while the Government of Bengal refused to divert funds from local and more pressing sanitary and medical projects to the isolation station, since the yellow fever threat remained small.[30] The end result was that the Government of India cancelled its order for the Government of Bengal to build a quarantine station, and six years after port isolation stations had been first proposed, no progress had been made.

A major obstacle to policy development was that the provincial governments did not feel the same sense of urgency that the Government of India did. Charles Bentley, the Sanitary Commissioner of Bengal, was not convinced that an isolation camp was necessary. He pointed out that the Government of Australia's recent attempt to enforce quarantine against the importation of influenza during the pandemic had failed. Australia had consequently abandoned quarantine in favor of systematic observation of persons infected or suspected with disease. Bentley thought that a systematized observation of crews of infected vessels would obviate the need for and be more efficient than quarantine, which was outdated, of doubtful efficacy, and involved unnecessary and detrimental delays to shipping and commerce. Furthermore, the introduction of yellow fever into Calcutta by sea was only a very remote possibility. Yellow fever would more likely enter Calcutta over land than by sea, and so the Government of Bengal did not see the point of funding an isolation station to the detriment of other public health projects.[31] The situation with the Government of Madras did not fare any better. The Government of India was not hopeful that Madras would be able to set up a quarantine station in the near future either. Given the reluctance of provincial health authorities and the financial hurdles to establishing an isolation station in Calcutta and Madras, the Government of India decided to concentrate attention on erecting a quarantine station in Bombay, which if necessary, could be designated the first port of call for yellow-fever-infected or -suspected ships arriving in India.[32]

The Government of India encountered several other difficulties in trying to implement its yellow fever strategy. Its policy had four main components: the protection of principal Indian ports from yellow fever infection; the elimination of Stegomyia mosquitoes at ports; the imposition of a strict system of modified quarantine; and support for precautions taken outside of India. The center's and provinces' financial situation had hindered the improvement of port sanitation. Stegomyia elimination had proven too time-consuming and difficult, since ridding ports of mosquitoes required an improved water supply, which officials hardly regarded as feasible within a reasonable timeframe. Other countries had criticized India's stringent quarantine rules for yellow fever, and there had been delays in coordinating precautions outside of India.[33]

At any rate, no such precautions, however elaborate, could fully eliminate the risk of yellow fever entering India, especially considering the inability of the government to respond to epidemics. The 1918–19 influenza pandemic had highlighted major defects in India's public health organization for addressing epidemic disease. There had been delayed notices of abnormal morbidity and mortality, a shortage of staff to investigate promptly the causes of abnormality, and a lack of arrangements to provide medical aid to those out of reach of hospitals and dispensaries.[34] Health officials experienced first-hand how ill-prepared India was to tackle a relatively short pandemic, let alone one that could become a permanent fixture in India's disease landscape. Officials emphasized that with the war over, India should waste no time in strengthening its health organization and in preparing against the possibility of 'what might prove to be an over-whelming disaster' if yellow fever entered the country.[35]

Given the inadequate state of India's disease preparedness and the realization that it could not continue to act as a renegade government, the Government of India decided to reconsider its yellow fever policies.

The Indian Port Health rules regarding yellow fever faced foreign opposition, and Britain had already denounced India's quarantine provisions. The Government of India believed that it would be hard pressed to justify its stance and to resist increasing pressure to relax its port rules unless it could show that every effort was being made to improve the sanitary conditions of ports, making them less vulnerable to yellow fever importation. The Government of India was anxious that a 'real attempt' should be made to grapple with the problem of port sanitation. If endemic yellow fever areas had been able to overcome their particular difficulties and had successfully reduced the Stegomyia population, then provincial governments in India, especially Bombay, should be able to as well.[36]

To reassess policy and discuss preventive measures against yellow fever importation, the Government of India assembled a Committee on Yellow Fever in February 1920.[37] The committee concluded that it was impracticable to separate the destruction of Stegomyia from the elimination of all mosquitoes. They recommended that the Government of India appoint a sanitary authority for each port, establish an organization in major seaport towns to control mosquitoes, contribute funds to these measures, and assume responsibility for any measures which local authorities failed to implement.[38] Although only five major ports existed in British India, there were approximately 200 small ports, 100 of which were in the Madras Presidency and 75 of which were in the Bombay Presidency. Some of the small ports had direct trading relationships with ports in East Africa, which in turn communicated with endemic areas in West Africa. These communications underscored the importance of taking adequate precautions at Indian ports.[39] The Committee also felt that the defense of India from the introduction of epidemic disease depended first and foremost on adequate sanitary surveillance of ships during their stopovers in infected ports with which India was in maritime communication. If the government could place absolute confidence in the effectiveness of this surveillance and if qualified medical personnel monitored all ship passengers en route, then measures undertaken at Indian ports of arrival could be considerably relaxed. The Committee reiterated the importance of implementing all necessary measures at the infected port of departure.[40]

A myopic Government of India chose to approve those committee recommendations that concerned only areas external to India's borders, namely foreign ports of departure, and to ignore suggestions for internal measures. Since the committee had argued that successful prevention depended on early notification by ports in America,

Africa, Australasia, and Asia that were in maritime communication with India, the Government of India asked the Secretary of State for India, Edwin Montagu, to introduce an international system of telegraphic notification for epidemic disease outbreaks in ports in communication with India.[41] The Government of India wanted Britain to set up a system of yellow fever notification with ports in infected areas that were either between the original ports of embarkation and India or in direct communication with India. S. P. James successfully argued on India's behalf that because yellow fever threatened the British Empire as a whole, addressing the issue required transimperial cooperation and coordination.[42] The British government agreed to set up the system if the Government of India would defray telegram costs, to which the Government of India agreed.[43]

India realized that a system of global outbreak vigilance could help identify potential harm and alert authorities to the need for intervention in order to avoid a situation from transforming into a domestic or international catastrophe. This system of notification, like the Singapore Far Eastern Epidemiological Bureau, aimed to detect, warn, and respond. Vigilance was a useful technique, which by attempting to prevent mass harm, allowed the co-existence with a danger when the alternative of prohibiting global trade and communication was neither possible nor desirable.[44]

The Government of India realized that notification could protect India only so much. It, therefore, asked the Secretary of State to coordinate an international effort to systematically eliminate yellow fever from endemic parts of West Africa. It also asked the Secretary to propose an agreement among governments threatened by yellow fever infection, which would outline the measures needed to render ports safe from yellow fever.[45] A skeptical F. J. Marchbank, the Under-Secretary of State for India, believed that it would be difficult to press for international action to eliminate Stegomyia in intermediate ports between India and areas of endemic yellow fever until comprehensive measures to rid Indian ports of the mosquito had been implemented, especially given the Yellow Fever Committee's report stating that existing arrangements in Indian ports were 'wholly inadequate' to prevent the introduction of epidemic disease.[46] Indeed, other than the notification system, the Government of India's requests for international efforts to control endemic yellow fever were not given effect.

By the late 1920s, no definitive measure of progress had been reached in the provinces, and very little significant public health legislation had even been passed by provincial legislatures since 1919. There were few, if any, measures to prevent the spread of disease between provinces, and interprovincial communication remained poor. League officials in Geneva and Singapore were more likely to obtain a quick perspective on the state of epidemic diseases in India than was the provincial sanitary official.[47] India still had no port quarantine stations in which to segregate infectious persons and their contacts, a situation that remained hampered by provincial officials' views that since there was no sanitary control over rail transport, it did not make sense to spend money on quarantine stations for ports alone. Moreover, they thought little chance existed for importing a disease not already endemic in India.

The Government of India rejected this logic in theory, if not in practice, when it agreed to the terms of the 1926 international sanitary convention and consented to consider provisioning quarantine stations in India.[48] The convention had two primary outcomes: it recognized the right of every country to subject arriving ships to medical inspection and quarantine based on the condition of the ship at the time of arrival and the disease history of the ship en route, and it obligated countries to provide adequate sanitary organization and equipment for at least one port on each seaboard and to send the Office International d'Hygiène Publique information regarding disease outbreaks at ports.[49]

Agreement was a convenient tactic for the Government of India; it provided greater flexibility in deciding to what extent it would abide by the convention. Signing a convention without ratification implied that states would in good faith try to respect the convention's provisions. Only ratification made states politically beholden to other parties to the convention. India tactically continued to delay ratification into the late 1930s because ratification would have entailed expensive and resource-intensive obligations in Indian ports. However, the Government of India intended to adhere to the 1926 convention as much as possible, including giving immediate notification of the first occurrence of notifiable infectious diseases.[50]

## Yellow fever takes to the air

The 1926 convention, although it aimed to provide protection against yellow fever, did not sufficiently address India's concerns about the importation of the disease. The convention really only dealt with the possibility of infection spreading via ships and did not take up the issue of the spread of yellow fever from endemic areas in West Africa to East Africa – a problem that health experts thought would only worsen with the completion of the transcontinental railway and with the growth of air traffic within Africa.[51] The convention proved even less useful because by the late 1920s, the possibility of yellow fever's entering India from East Asia via sea had diminished due to the control of yellow fever in the Americas, and the East had also remained free from yellow fever despite favorable climatic, vector, and sanitary conditions.

Given the decreasing possibility of infection coming from the East via ships and the increasing likelihood of it entering from Africa, the Government of India shifted its attention away from the Americas to yellow fever regions in Africa.[52] A journey from an endemic area in western Africa to a port on the eastern African coast could be completed by rail, car, or plane within the incubation period of the disease.[53] The distance between harbors on the east coast of Africa to ports on the west coast of India required only eight days' sailing time.[54] Air travel was increasing not only within Africa but also between Africa and India; the first Imperial Airways flight between Egypt and India took place in December 1926. The intention of sanitary authorities in West Africa to pass the responsibility and expense of yellow fever containment to health authorities at destination ports only compounded the Government of India's fears of not being able to protect India.[55]

Fortunately for India, the League and the OIHP were also studying the problem. In 1930 the Eastern Bureau of the League of Nations passed a resolution recommending that all Eastern countries prohibit airplane communication with yellow-fever-infected or -suspected areas until measures outlined by the OIHP were implemented.[56] The League worked with governments and in cooperation with the International Commission for Air Navigation to review the risks of transmitting infectious diseases by aircraft and to determine what international action was required. At the League of Nations conference in Cape Town in 1932, at which Graham was present, delegates, having not yet contemplated that the Anglo-Egyptian Sudan could already be infected, agreed that East Africa should be protected as a buffer zone for India. The Rockefeller Foundation's Health Division moved to extend its yellow fever immunity survey into British East and Southern Africa with the help of the Colonial Office.[57] To complement the League's efforts, the OIHP began work to introduce a convention to control 'aerial navigation', which would consider aerodromes as special sanitary enclaves and subject to the strictest mosquito controls.[58] The British delegate to the conference, George Buchanan, had been able to argue successfully that since aerodromes were easily safeguarded from infection, the onus of disease control lay largely with the exporting country. This resulted in the 1933 International Sanitary Convention for Aerial Navigation stipulating that before governments of endemic areas could permit international travel, they would need to build anti-amaryl, or yellow-fever-proof, aerodromes.[59]

By the time of the aerial convention, international public health experts were viewing yellow fever as the easiest of all the infectious diseases to control. Some said that the convention's provisions went beyond what was indicated or necessary. Medical research had advanced epidemiological understandings of yellow fever, rendering it more easily contained. Ironically, S. P. James, who had previously called for increased vigilance by the Government of India and who had since become advisor to the Ministry of Health for Britain and President of the Yellow Fever Commission of the OIHP, viewed the Sanitary Convention for Aerial Navigation as alarmist.[60]

The Government of India, on the other hand, did not think the provisions of the 1933 International Sanitary Convention for Aerial Navigation were sufficient to protect India. It worried that passengers flying from infected places in West Africa could reach coastal ports in East Africa, board a vessel there, and arrive in Bombay within nine days of having left an infected area. The Bombay Port Health Officer had already reported the arrival in Bombay of a passenger from Wau, a town in East Africa where yellow fever cases had appeared.[61] Even though there were no direct lines of air traffic between yellow-fever-infected zones in West Africa and uninfected areas in East Africa, air transit times had decreased over the previous year.[62] Furthermore, evidence had shown that mosquitoes could survive airplane travel over 1250 miles and might remain on aircraft during stopovers at aerodromes.[63]

Public health officials in India reasoned that these circumstances necessitated greater precision in the convention's classification of infected and non-infected areas. According to the convention, non-endemic areas in Africa were classed differently than endemic areas and thus fell under different regulations. The Government of India thought that 'silent areas', in which the presence of the virus had been detected although no clinical cases had presented, should be classed as infected if they demonstrated positive mouse protection tests.[64] When the 1933 convention had been framed, the only infection possibility envisioned lay in western Africa, the endemic part of the continent. Since then, discoveries of silent areas of yellow fever infection had been detected in Egypt, the Anglo-Egyptian Sudan, and Uganda, which meant that the potential area of infection in Africa was much larger than health authorities had imagined.[65] Since the convention did not recognize silent areas as infected, medical researchers and scientists in India found the convention precautions inadequate. Officials in India thought that delimiting areas of clinical and silent infection was crucial to aerodrome protection, and they objected to the convention's stipulation that anti-amaryl aerodromes and the application of other requirements were obligatory only in areas of clinical infection and not in silent areas.[66]

The exemption of silent areas from convention provisions weakened the Government of India's public health defense. The Public Health Commissioner, G. G. Jolly, argued that, until the degree and extent of infection in silent areas was ascertained and the limits of those areas demarcated, it was only prudent that India should view silent areas as potential sources of yellow fever infection and take the necessary precautionary steps.[67] Consequently, the Government of India refused to accept less stringent standards than those provided for in the 1933 convention and, in fact, argued for special allowances to implement stricter precautions.[68] J. W. D. Megaw, the Government of India's delegate to the OIHP, obtained modification of the convention to allow the Government of India to refuse or prohibit the landing of any airplanes coming from yellow-fever-infected countries at the Karachi aerodrome, which was the only point of debarkation for flights from Africa.[69] Having obtained international sanction, the Government of India issued notifications in 1936 under the Epidemic Diseases Act prohibiting entry of any aircraft that had originated from or alighted in an endemic or silent area, except those aircrafts that had obtained a certificate of disinsecticization at Alexandria or Cairo.[70] The government appointed a health officer to monitor sanitary conditions at the Karachi aerodrome and to bring it into compliance with anti-amaryl specifications. The Government of India later decided to 'disinsectize' all aircraft coming from suspected or infected areas as general practice.[71] The Government of India also went two steps further than the convention's provision, which stipulated that a person traveling from an infected area could not enter a country free from yellow fever for at least six days. The government extended the waiting period to nine days (to cover the six-day incubation period and three days of infectivity) and expanded the regulation to include silent areas as well.[72] India worked out an arrangement with Egypt, so that the Director of the Egyptian Quarantine Board would telegraph the Airport Health Officer at Karachi to warn him if a passenger coming from a suspected or infected area in Africa had proceeded to India within the nine-day waiting period.[73] Passengers and crew members that had been in an endemic or silent area had to show proof of inoculation or of acquired immunity from an attack of the disease or spend the entire incubation period in a non-infected area before arriving in India.[74] While India did what it could to protect Karachi, the Director-General of the Indian

Medical Service, Cuthbert Sprawson, realized that '"reliance on this one barrier on the Indian side alone is not enough; it is necessary also that the utmost possible be done on the African side of the gate to prevent infective material approaching it"'.[75] India, however, was powerless to influence prevention measures in Africa, so the Government of India focused on strict sanitary control of its airspace and aerodromes.

The Government of India justified this rigid and defensive stance in the face of increasing risk. The Rockefeller Foundation's yellow fever immunity survey had found a much wider distribution of yellow fever in Africa than had been previously suspected – from the Senegalese coast eastwards for 3300 miles to the upper White Nile in the Anglo-Egyptian Sudan.[76] In 1937, a new air route had also opened between Lagos and Khartoum, which then linked up with the Cape Town-Cairo route, thereby increasing the possibility of carrying infected people and mosquitoes from endemic areas in West and Central Africa to East Africa and on to India within yellow fever's incubation period.[77] Furthermore, an endemic type of yellow fever had appeared in districts in Africa where no Stegomyia had been found. A new type of yellow fever called jungle yellow fever (later termed sylvatic) had also been discovered; it existed in epizootic form and could occasionally be passed from monkeys to humans, especially to agricultural laborers, by the *Stegomyia fasciata* mosquito. The blood of monkeys, unlike human blood which remained infective for only the first three days of the disease, was infective for the duration of the disease. For India this meant a potentially vast monkey reservoir if this variant found its way into the country.[78] If monkeys became a reservoir, it might not be possible to eradicate yellow fever for generations, if ever.[79] Furthermore, experiments conducted on rhesus monkeys at the Institute for Tropical Hygiene in Amsterdam had demonstrated that Indian mosquitoes could be as vector competent as their American brethren. Climatic conditions in India would only favor the development of mosquito infectivity. To make the Government of India even more uneasy, four non-aegypti species prevalent in India had been experimentally shown to transmit the yellow fever virus to monkeys.[80] The extent of animal reservoirs and mosquito vectors in India and the risk of yellow fever spreading if it took hold meant that containment of the disease would be incredibly difficult.

## Conclusion

Despite the Government of India's inability to achieve proper port sanitation and mosquito eradication, India escaped yellow fever's wrath (and has remained free of yellow fever to this day), although this probably had more to do with the epidemiology of yellow fever than with the government policies of the 1920s and 1930s. The Government of India could not have known that India would be spared. Not only that, it had no means of making risk comparisons with any other extant disease in India. In analyzing diverse risk scenarios, comparisons could have provided the Government of India reassurance or perspective on a particular risk.[81] In the case of yellow fever, the government could only make external comparisons based on the consequences of yellow fever epidemics in other countries, which, however, had different epidemiological conditions.

The government's fear of a very real and, for India, new public health risk created an atmosphere in which reactionary public health protectionism was not only sanctioned but justified, at least in the Government of India's eyes. The risk of this high-consequence disease outbreak oriented government action and policies, which took into account the cost and benefit of inaction as well. Risk management was challenged both by the uncertainty of experiencing an unprecedented disaster and the need to deter the 'temporal and spatial extent of the consequences' of an eventual epidemic.[82]

In the end, the government's political will and financial means led it to favor precaution over preparedness. Whereas precaution seeks to avoid an uncontrollable risk, preparedness assumes the eventual realization of that risk and thus develops a response mechanism and thresholds to mitigate rather than to prevent.[83] The government also operated on a principle of precaution because the risk was not clearly known and security could not be achieved. Precaution implied a public policy in which risks were viewed from a moral perspective, since strong scientific proof did not exist but the potential for irreversible harm did. Prevention was an essential tool used by the Government of India to deal with scientific uncertainty in the face of potential hazard.[84] It also conveniently aligned with the government's policy of least financial resistance, political and financial constraints having precluded a policy of mitigation.

The government's concern to protect India was further heightened by advances in understanding yellow fever's epidemiology, which underscored India's vulnerability to the disease. Laboratory technology had identified a larger area of yellow fever infection that extended beyond the zone of endemicity, helping define the disease at the molecular level instead of at a clinical level.[85]

If fear and perception of epidemiological risk created the conditions for articulating a yellow fever policy, then factors such as Devolution, colonial finance, resource availability, and logistical practicability structured the actual content. Safeguarding India required the institution of protective measures in Africa, sanitary control of India's airspaces, and cooperation between the central and provincial governments to improve port sanitation. Unfortunately, only one of these proved viable. The Government of India was powerless to impact the first, did what it could to undertake the second, and after Devolution had less jurisdiction to affect the third. The Government of India's sense of urgency contrasted with its ability, or lack thereof, to take decisive action within India's borders. Under the new 1935 Constitution, the Federal Legislature continued to relinquish power to legislate on provincial public health and sanitation. Although port quarantine was a central subject, sanitation came under provincial purview, and since yellow fever had not yet appeared in India, the Government of India could not force provincial governments, under the Epidemic Diseases Act, to improve port health and eliminate Stegomyia breeding grounds. The Government of India also declined to assist provinces by providing funds to ready ports, which then increased tensions between central and provincial health officials.[86] The attitude of some officials in the public health establishment only exacerbated the situation. They believed 'expenditure on the prevention of the introduction of infection is not justified when infection already exists. . . . [The] endemicity of many infectious diseases in India has undoubtedly been at the bottom of the present unpreparedness of India to meet the yellow fever danger'.[87] The government thus relied on protectionist, severe, and sometimes outdated disease control measures to police its borders, even if it meant jeopardizing international health conventions and aggravating officials back in London.

With the provinces unable to fund adequate port sanitation and the Government of India powerless to enforce sanitary measures and financially unwilling to provide assistance to the provinces, the Government of India decided to concentrate on the sanitary enclaves it could control – ships, airplanes, and even laboratories. Ships and passengers arriving from yellow-fever-infected or -suspected areas were closely monitored. As the air became the 'new ocean', providing a frighteningly quick link between India and endemic and infected areas of Africa, the Government of India turned its attention to regulating airplanes and aerodromes.[88] By the outbreak of World War II, air transport had placed India within a 24–36-hour journey of yellow fever endemic areas in Africa. Due to the considerable expansion of aerial communications in the 1930s and the extension of war to Africa, the risk of introducing yellow fever into India had increased.[89] British India and its tropical neighbors were also keen to avoid risking accidental infection of mosquitoes or of people through laboratory experiments. India joined the Federated Malay States, Dutch East Indies, and Straits Settlements in enacting legislation to prohibit the possession of amaryl virus, even for research purposes, in their territories. Laboratories under the Government of India's control, including the IRFA, had already been prohibited from using the yellow fever virus in experiments, and the government moved to extend this ban to private laboratories as well. In 1931, the Governments of Bombay, Bengal, and Madras approved the Government of India's total prohibition of the importation of yellow fever virus into India.[90]

India's policy advantage was that yellow fever's 'geoepidemiological trajectory' did not include the European continent, enabling the Government of India to maneuver within the international public health arena in a way that better protected India's interests. Even the British delegate to the 1911–12 International Sanitary Conference, R. W. Johnstone, Medical Officer of the Local Government Board, suggested that yellow fever had no place in a 'European' conference, since the risk of transporting the disease from the Americas to Europe was negligible.[91] The lack of an epidemiological relationship between Europe and India allowed India more room to implement stringent precautionary measures in light of the public health establishment's limited ability to respond should yellow fever enter India. At international conferences, the Government of India emphasized that India had a unique epidemiological position, and thus required policies more rigorous than Europe's. India's yellow fever policies conflicted with international scientific opinion and violated both long-standing and contemporary international public health agreements, but the Government of India was allowed this discretion. International indulgence enabled the Government of India to succeed in two areas: it implemented stringent, even if controversial, quarantine regulations, which aimed to protect India's ports, and it established strict rules of entry for airplanes and passengers arriving at the Karachi aerodrome.

Yellow fever was a threat from the outside, and the Government of India instituted precautionary measures, however minimal, to prevent the disease from entering India. In the same way that crises had driven the Government of India to act decisively to prevent plague and cholera from spreading to Europe, the epidemiological risk posed by yellow fever entry structured the Government of India's yellow fever prevention strategy. However, there were important differences in the motivations underlying the government's decisions. Whereas the protection of trade and commerce was a primary factor in shaping the Government of India's policies toward the control of epidemic plague and cholera, the protection of India was the main objective in its yellow fever policies. Unlike cholera, plague, or malaria, which had been devastating India for hundreds of years, officials could not calculate, let alone imagine, with any certainty what type of and how great an impact yellow fever would have should it infect India.

Like the Government of India's malaria policy, yellow fever policy was also 'cheap policy'. It required less expenditure than a far-reaching strategy that would have included mosquito-proofing major ports. Ships could be quarantined and airplanes turned away with relatively little expense or inconvenience to the Government of India. As a result, very few of the Yellow Fever Committee's recommendations had been implemented by World War II, and India remained in the same unprepared state as it had been in the 1920s.[92] Given the circumstances in which it found itself, the Government of India chose to implement stopgap measures, which, while criticized by public health authorities in Europe and India, reassured the Government of India nonetheless. Although 30 years of international experience had proven the possibility of successfully and rapidly eliminating yellow fever, this did nothing to alter the Government of India's perception of risk, precisely because the government did not have the financial means or human resources necessary to control the disease if it did enter the country.

# 5
# Disease as Prism

The Government of India made certain choices regarding four deadly infectious diseases, and for several reasons, the health of colonial subjects rarely entered into discussions of infectious disease policy at the central level. The Government of India's approach to health policy was largely instrumental, considering international and domestic political relationships and economic impact primarily and the well-being of Indians only secondarily, if at all. To understand why the government adopted this approach, it is necessary to look beyond a metropole-colony dialectic and examine how international health organizations and policies influenced and were influenced by governments and health officials in India and to understand the role that state perceptions of risk played in health policy decisions.

Health policy in India was influenced by more than the metropole or the colony or even the imperatives of the imperial system. India was simultaneously a sub-imperial node within a larger webbed empire and an imperial center from which people, ideas, and goods radiated outwards and transformed different parts of the world.[1] Between 1890 and 1920, this India-centered sub-imperial system reached its peak development.[2] Its position in a global economy and its centrality to the British Empire meant that India formed a critical part of an epidemiological web spanning East Asia, Oceania, Europe, the Middle East, and Africa. Viewing India in a web where peripheries acted as centers of other webs permits the removal of the center–periphery straitjacket and an understanding of how locations, individuals, and institutions at the periphery of one web formed the center of another.[3]

Within these interconnections, India held a prominent position in the international health community – a subject largely unexplored – where India was respected for its medical research and medical opinions and feared for its geoepidemiological position. For doctors and scientists in the colonies, research provided a legitimate means to make the periphery relevant by enabling their participation in a global community of investigators engaged in discovering knowledge that could be of use to the empire.[4] Public health officials in India, however marginalized by their imperial colleagues at the London and Liverpool Schools of Tropical Medicine, played a central role in international discussions about tropical diseases into the 1930s. By participating and constructing intellectual networks, which radiated out of India into the larger British Empire and around the world, medical researchers in India gained legitimacy both within and without India. This status provided further impetus to undertake disease investigations and establish research institutions and public health schools in India. Simultaneously, European study visits to the colonies contributed to the development of medical science globally.[5] India was not simply an experimental laboratory but also a site for scientific and intellectual breakthroughs.

India's position within this international public health arena shows that India's sovereignty was more fluid than has often been supposed. Tensions among different loci of health authority were heightened as a result of this fluidity, affecting policy outcomes, both within India and internationally.[6] India's liability to an international health community made it somewhat independent of Britain's will in the public health domain, while the political processes involved in negotiating international conventions contributed to the formation and evolution of India's semi-sovereign status within international public health.[7]

As India's seat at the diplomatic table became more secure, it became increasingly involved in public health discussions at the international level and in investigating plague, cholera, malaria, and other communicable diseases. Public health officials became more invested in research and in attending and hosting international and domestic public health conferences, even if the practical application of that research and the recommendations coming out of those conferences were rarely implemented to improve the health of the Indian population.

Indian research agendas were tied to the Government of India's desire to carve out a niche in the international public health arena and contribute to the growth of scientific knowledge. India's institutions advanced the League of Nations' health work on malaria, cholera, plague, leprosy, kala-azar, hookworm, and nutrition. In turn, the League provided a forum within which the Government of India could achieve prominence in tropical disease research.[8] Between the 1890s and World War II, India had become a center of medical achievement, a position that is often glossed over in favor of critiques of the colonial state's public health policies. India's cinchona production led to a drop in the price of quinine in the late 1800s, thus allowing its sale to be extended to the poor, and India became the first country to sell a quinine dose at the smallest coin in use. Ross discovered the mode of transmission of malaria; Haffkine developed anti-plague and anti-cholera inoculations; and the Indian Plague Commission elucidated the role of rats and rat fleas in spreading plague.[9] This list is not comprehensive, but it illustrates that during this period, India's contribution to tropical disease research and treatment reached new levels and obtained international acclaim.

As this account reveals, internationalization was not simply a process of diffusion, with international organizations and transnational networks gradually extending their reach from the West to Asia and Africa, nor was the globalization of Western medicine and medical institutions solely a one-dimensional imperialist project.[10] This analysis of epidemic disease policies suggests an alternative to both of these narratives – India's disease landscape put Anglo-Indian and later Indian medical men at the center of the development of international health organizations. Their participation in these international conferences and organizations aimed to establish British and Anglo-Indian scientific authority within an international arena, not to develop further means by which to control a colonized population. Policy discussions regarding plague, cholera, malaria, and yellow fever show a remarkable lack of desire to colonize, let alone cure, the Indian body, and they underscore the shortcomings of colonial governmentality. Policy for these diseases was largely driven by externalities to the sufferers.

## The role of risk in infectious disease policy

The juxtaposition of the histories of policy development for cholera, plague, malaria, and yellow fever underscores how the biological realities of disease interacted with political regimes, international communities, and economic, cultural, and social factors to influence the final form of policy in British India. Different diseases elicited different responses from governments and other stakeholders. These reactions reveal that, ultimately, the determination of and response to different kinds, degrees, and locations of risks and the delineation of potential consequences for populations are political decisions. The values and assumptions justifying decisions and the rationalizations provided to the public are based neither on law nor entirely on scientific reasoning.[11]

Governments and health officials around the world were continually haunted by historical memories of cholera, plague, and later, influenza pandemics. State fears of a repeat pandemic amplified perceptions of risk.[12] When plague erupted in Bombay in 1896, countries from the Americas to Europe were quick to institute quarantines and embargoes against India. The 1897 Venice convention laid out an internationally accepted protocol that then allowed these restrictions to be lifted and international commerce and pilgrimage to continue. The subsequent shift in international reaction to plague in India from one of alarm to one of risk management did not mean that plague no longer posed a threat to Europe. Plague became endemic to India, making the country the main reservoir of plague in the world. The evolution of international responses to plague in India shows that state understandings of public health risk are not absolute. They are subject to value-laden judgments that take into account particular political, economic, and cultural conditions and the hypothesized reactions to policy decisions that certain stakeholders and populations will have.

The perception of risk fundamentally shaped the Government of India's approach to policy-making on infectious disease. The type and degree of risk a disease posed to India depended largely on whether that disease could become epidemic and cross India's borders. Risk was conceptualized in terms of a disease's effects on commerce and health and in terms of how a particular response would affect political and economic relationships within India, throughout the British Empire, and internationally. The identification of risk then led to a recognized need to create policy. While risk was not quantified, there still existed an implied disaster threshold – a point beyond which misfortune would have been experienced as disaster. The threshold varied depending on the particular situation and on which side of the risk scenario officials sat.[13] For the Government of India, risk calculations equaled qualitative perceptions, and the threshold varied by disease. The response threshold for epidemic plague and cholera was relatively low, for malaria relatively high, and for yellow fever, practically zero. The appearance of any cases of yellow fever in India would have constituted disaster based on health officials' and policymakers' understandings of the disease's epidemiology and India's ecology.

Sometimes health officials disagreed over the extent of risk and the mode through which risk was transmitted, which complicated the policy-making process. Robert Koch critiqued the 1893 and 1894 international sanitary conventions, stating that closing or controlling the Persian Gulf and Red Sea served no purpose in preventing the ingress of cholera, since cholera had previously spread overland from Asian steppes to Southern Russia and into Europe. Europe, however, remained fixated on policing maritime routes.[14] There was a perception on the part of European states that cholera *could* spread to Europe via ships carrying infected passengers. Quarantine, regardless of its disputed efficacy, was easier to implement and police than a land-based sanitary cordon across thousands of miles of borders and was also more reassuring to European populations.

The international regulation of India became the consequence of these global perceptions of risk. Both international and Indian perceptions of risk were fundamentally informed by one critical circumstance: the epidemiological relationship between infected and uninfected areas within a disease's epidemiological trajectory. The geographic, economic, and political distribution of risk from disease was not equal across regions or countries; as a result, perceptions of risk varied with that distribution.[15] Whether traversed by ship, caravan, or plane, the distance of an uninfected country from an epidemic's start point or from an endemic zone determined the degree of protection a state felt and the amount of room and time a state had to maneuver and learn from others' experiences. In reaction to cholera and plague outbreaks in India, Britain, by virtue of its northerly position detached from the Continent, was afforded a degree of comfort not enjoyed by Russia, Turkey, or France, which were more closely connected to India through land and maritime routes. British officials knew that a case of cholera or plague would have presented in the time it took a ship to sail from India to Britain. Other countries did not necessarily have that luxury. As a result, a roughly north–south split in quarantine preference developed in Europe. Southern Europe, being closer to the 'source' of cholera and plague, favored stricter quarantine.[16]

The epidemiological relationship between Europe and India with respect to cholera and plague created economic and political risks for India. European actions and fears negatively affected trade and labor flows. Politically, epidemic disease control pitted India against other states and jeopardized its relationship with Indian pilgrims. In order to keep people and goods moving in a manner acceptable to merchants and to Indian governments, the colonial state had to implement policies that would appease European states and prevent further damage to India's economy and peoples, while still ensuring that Indian exigencies were met or could safely be deferred, as in the case of the pilgrimage ban.

The epidemiological relationship functioned in the opposite direction as well. In the case of yellow fever, India's proximity to Africa and the increasing speed of maritime and air travel over the first half of the twentieth century created much anxiety within the Government of India and among health officials. India's buffer zone was diminishing quickly as immunity surveys discovered a much larger zone of infection in Africa than had previously been thought and as air travel brought Africa even closer to points of entry into the subcontinent.

Malaria presented an entirely different case. The lack of an epidemiological relationship between India and other states meant that there was no external pressure for India to address the disease comprehensively or systematically, nor was there an internal sense of urgency to try to reduce malaria mortality and morbidity, even though officials unanimously recognized the economic and health consequences of malaria's wrath. The absence of an international stimulus gave the Government of India more flexibility in its health policies. With regard to malaria, this combined with the relegation of public health matters to provincial and local governments to translate into sporadic and superficial attempts to control the disease.

## Structuring policy

If the perception of risk, either by foreign states or by India, created the requisite conditions for creating policy, it also served as a constant referent in the formulation of policy. Government perceptions of the political, economic, and public health risks resulting from the threat of infectious disease and the presumed consequences of actions taken or not taken to prevent and control disease fundamentally shaped policy development on plague, cholera, malaria, and yellow fever. The types of risk, the way in which those risks were perceived, and the ramifications of addressing or ignoring those risks varied by disease, political climates, economic priorities, and the amount of political will to control disease.

Within state conceptions of risk, several factors combined to structure the content of policy. The significance of those factors changed over time and varied by disease. Epidemiology and etiological understandings affected what public health measures the government could implement. The epidemiological forms of plague and cholera acted as critical variables in the determination of policies because the perception of risk for the epidemic versus the endemic differed. For epidemic disease, political and economic risks motivated the Government of India to employ decisive measures. At the same time, epidemiological and etiological data influenced actual measures on the ground. The discovery of mosquitoes and rats as vectors of malaria and plague, respectively, added new targets for disease control and allowed a relaxation of pilgrimage restrictions, while better understandings of the cholera vibrio redoubled efforts to monitor domestic religious festivals. The determination of yellow fever's incubation period influenced the length of quarantine for travelers from infected regions.

In conjunction with scientific discoveries, the development of new medical technologies also altered, and generally expanded, the Government of India's arsenal of health measures. Anti-plague and anti-cholera inoculations became central elements of the government's strategy to address these diseases. The government's reliance on immunization was predicated on a recognition that prior sanitary strategies, to the limited extent they existed, had failed to control dangerous individuals and practices. Inoculation, which was far easier to implement than widespread sanitary improvements, tried to reduce the risk of disease among target populations and operated on a logic that made individual actions less relevant.[17]

If epidemiology and medical technologies affected what options were available to governments, political relationships, both international and domestic, shaped which options were actually implemented. The International Sanitary Conferences were not simply straightforward international health discussions. They were also a forum in which debates on policies were as inextricably tied to negotiations of the boundary between nationalism and internationalism as they were to determining the role of science in diplomacy and the relationship between the West and the Orient.[18] Disease diplomacy implicated both imperial and international relationships and, therefore, reflected balance-of-power politics. The globalization of health issues and disease control threatened national sovereignty in each state's ability to decide what was best for itself. As a result, assertions of national sovereignty and attempts to protect it were never far from the minds and actions of delegates to the conferences, even when those actions could and did hinder disease control internationally.[19] Delegates had to strike a balance between protecting national and imperial interests and achieving international health cooperation in order to control epidemic disease and ensure smooth trade flows among empires and states.

To prevent governments from imposing severe trade restrictions, Britain and India had to accept limitations on their sovereignties. The very goal of self-preservation required an acceptance of obligations to an international community.[20] Common interest and coercion, in the form of trade embargoes and British officials pulling imperial rank, had converged in the international health arena to achieve India's obedience.[21] Without a supranational power to enforce international health agreements or coordinate responses to epidemic disease, European states relied on economic threats to achieve India's compliance to regulations set forth in the international health conventions ratified between 1892 and 1897.[22] The Government of India recognized that its violation of the conferences' multilateral international agreements would incur a negative reaction from participating countries.[23]

There was one important exception to the privileging of foreign relations, however, and that was yellow fever policy. Europe did not risk yellow fever becoming endemic because the continent lay outside of the latitudinal band of the disease. This allowed the Government of India's officials, who believed that the advantages of violation outweighed the advantages of obedience, to disregard law and international consensus.[24] This instance shows that the lack of an epidemiological relationship between India and Europe permitted India to bend international convention to its needs without any lasting damage to its political or economic relationships and with the added benefit of asserting its diplomatic independence in the international public health arena and within the British Empire.

The Government of India's risk-driven policy became intertwined with a sense of public health responsibility toward the world. The same factors that influenced policy decisions also pushed India to become more responsible for its diseases, at least to a certain extent and with that extent varying by disease, and to become more active in disease research. India, with its large population, its global trading position, and its incidence of plague, cholera and smallpox, created unease among the public health authorities in other countries.[25] J. D. Graham, Public Health Commissioner, noted how 'rapidly the international side of our work and of our obligations in the domain of public health is opening up, and how essential it is for us to have an adequate appreciation of what is going on throughout the world and to be able to take our place in line with other nations when opportunity presents itself either in regard to new developments or in regard to the organisation of work on international lines'.[26] The India Office also pressured the Government of India to participate fully in international health discussions. When India refused to send its Public Health Commissioner to represent India on the Permanent Committee of the International Bureau of Public Health, the India Office fired back that the activities of the League's Health Organisation and the establishment of the Epidemiological Bureau in Singapore had directed greater attention to health administration and

port sanitation in India. International health diplomacy and the prestige and material interests of India necessitated that a qualified health
authority from India be present at important international meetings.[27]
Officials in both Britain and India had come to realize that by virtue of
India's disease landscape and central role in international trade, India
had an international responsibility to control epidemic disease and to
take her position within a community of nations.[28]

India's international health mandate also stemmed in part from a
desire among health officials and scientists to see India self-sufficient
in terms of research and disease control. Increasing international attention on India in the early 1890s coincided with (or maybe caused)
public health officials' move for India to become an authority on
tropical disease. India needed to find internal resources to address its
disease problems. There was a sense, imbued with pride and a competi-
tive, imperialist spirit, that India had much to contribute to international health in terms of research, resources, and practical expertise.
The Government of India envisioned an important place for India in
the international health arena. The Office International d'Hygiène
Publique and the League of Nations provided forums within which the
Government of India could establish its prominence in tropical disease
research. To that end, research as public health policy and research as
legitimacy became intimately intertwined.

The fact that the League requested participation of expert health
officials from India on its commissions attests to the success of the
Government of India's project. The League was at its core a regionally
oriented organization; the primary goal of the Health Organisation,
like the League itself, was European security, both internal and exter-
nal and inclusive of colonial possessions and mandated territories.[29]
Officials in India and in other Asian lands helped broaden the Health
Organisation's purview, starting with the creation of the Singapore
Bureau and expanding to include tours to study the health problems of
Asian countries.

The issue of public health responsibility domestically was not as relevant to policy decisions as India's public health obligations internationally. The Government of India's approach to fighting infectious disease
was largely instrumental: it primarily served commercial and political
purposes and was only marginally concerned with the intrinsic health
of its colonial subjects. Advances in disease research and the containment of disease within India's borders could be paraded and applauded
in front of international congresses and at the League's Health
Organisation, cementing the centrality of British India to progress in

tropical disease research and in global disease control. However, the state of health between India's coasts, as long as it did not threaten other nations, was not of international political importance, even as non-governmental organizations, such as the Rockefeller Foundation and the Indian Red Cross Society, became increasingly involved in bettering the health of India's millions. Research afforded India bragging rights, while its mortality and morbidity statistics remained shameful. The number of investigations conducted in India on cholera, plague, and malaria, India's most debilitating disease, clearly shows that the Government of India recognized the importance of these diseases to India's present and future. That recognition, however, did not lead to good policy decisions regarding endemic disease. The Public Health Commissioner with the Government of India asked in 1922: 'What is the value of a human life in India? Again what is the economic loss due to sickness?' These questions lay at the heart of public health policy in India.[30] By 'value' the commissioner seemed to refer to the economic and not the moral value of life. Medical research had become an 'instrument' of the colonial state as a source of legitimacy, rather than a means by which the 'idiom' of development and improved health could be achieved.[31] This way of framing is still current in some of today's public health analyses, in which a human capital approach views health investments in their relation to improving economic productivity.[32] What is striking is not so much the commissioner's questions but the answer. The economic loss was immense. Yet this was not enough to overcome other significant and sometimes extremely entrenched obstacles to improving the health status of Indians.

Tension between recognition of the need to prevent and treat infectious disease and the Government of India's ability or willingness to control disease pervaded public health policy discussions and influenced policy decisions throughout this entire period. It played out in different ways for cholera, plague, malaria, and yellow fever. With respect to cholera and plague, European fear coincided with increasingly powerful international public health diplomacy to force India to contain epidemic outbreaks. The Government of India reconciled the need to control disease with its limitations by privileging the control of epidemic outbreaks over the reduction of endemic disease, by using its epidemiological exceptionalism to maneuver in the public health arena, and by focusing on sanitary enclaves to address both epidemic and endemic disease. This last approach precluded the delineation of an all-India strategy that could be administered or overseen by the central government. Instead, the Government of India directed its most potent disease control and

prevention measures at specific sites, including religious fairs, ships, laza-
rettos, tea gardens, ports, and research institutes. These were physical,
delimited spaces that the government could target, but they were also
areas that were either lucrative or under international scrutiny.

Malaria also suffered from an enclavist approach directed at construction and rail works and tea estates. From the point of view of
epidemiological and economic risk, the Government of India's policies
toward malaria control are inexplicable. But if one takes into account
a growing concern among officials in the Colonial Office and within
the Government of India to maintain India's credit-worthiness and its
ability to meet its external financial obligations, then one can see why
the Government of India was reluctant to allocate significant funds to
controlling malaria or any other endemic disease.[33] Fiscal responsibility
seemed to guide malaria policy more than economic growth. The fiscal
obligations that India had to meet set an upper limit to public health
expenditures, with colonial ideologies and practical difficulties further
lowering that limit.

Malaria did not benefit from politicking or perceptions of political
risk either. The politics of collaboration did not affect malaria policy,
as they had done with plague and cholera control, and the prevention
of the disease did not gain much material support by the devolution of
public health to provincial and local governments. There was no strong
international or domestic political incentive to address malaria in a
more comprehensive manner, so the Government of India continued to
focus on short-term and inadequate solutions. Medical opinion, however fractured, informed policy but was subordinate to fiscal constraints
and domestic political factors.[34] Officials in India themselves realized
that while India had created a wealth of knowledge and implemented
measures to address epidemic plague and cholera, it had yet to take
advantage of the malaria research it had so proudly produced.[35] As a
result, by the 1930s, malaria was killing at least three times as many
people annually as cholera, plague, and smallpox combined.

Yellow fever policy, in contradistinction to that for the other three
diseases, was primarily driven by the desire for domestic epidemiological security. In policy discussions, it was not clearly articulated what the
potential consequences would be because apart from thinking that it
would be catastrophic, health officials could not concretely imagine the
extent or type of havoc yellow fever would wreak in India. It was precisely this fear of the unknown that explains the Government of India's
reaction. With respect to yellow fever, India was 'virgin soil'. As history
has repeatedly demonstrated, virgin soil populations are particularly

vulnerable to outside infectious agents and have been frequently devastated upon the import of new bacterial and viral diseases. While the
sense of urgency to prevent yellow fever from entering the country was
intense, the Government of India was still subject to fiscal, political,
and resource constraints, which limited the types of policies it could
implement and created support for an enclavist approach.

The control and prevention of yellow fever, as of malaria, did not
impinge on India's domestic and international political relations in the
way that cholera and plague epidemics did. So although the perception
of risk to health and economy was great for both malaria and yellow
fever, that risk was not accompanied by international pressure for the
Government of India to manage India's diseases effectively. Precisely
because of the absence of European concern with India's yellow fever
situation, the Government of India was able to put in place stringent
precautions at ports and aerodromes to protect the country from yellow fever. The contrast between approaches to malaria and yellow fever
control shows that epidemics that had international ramifications
galvanized the Government of India to a much greater extent than did
endemic or epidemic diseases which did not threaten to cross borders.

Given its financial, logistical, administrative, medical, and attitudinal
constraints, the Government of India chose to focus its prevention,
treatment, and containment measures in situations of political or economic emergency and at points of control where it could be the most
effective. Infectious disease policy in India was an exercise in acute crisis
management.[36] The prevention of epidemic cholera and plague required
the ordering and cleansing of internal and external spaces. Fairs, bazaars,
and festivals, as places of mass congregation, needed sanitary regulation to avoid cholera outbreaks, while dwellings needed disinfecting to
prevent plague.[37] Borders were also critical fronts in crisis management.
They were conceptualized by delegates from India and Europe as 'semi-
permeable membranes' which could be open for colonial and commercial enterprises and certain people and closed for others. Inherent in this
conceptualization was the categorization of groups of 'border crossers' as
dangerous and not dangerous – migrants, laborers, gypsies, and pilgrims
falling under the former category, which later expanded to include travelers from yellow fever endemic zones in Africa.[38]

## Obstacles to policy development

While a constellation of factors structured policy discussions and outcomes, there were also a number of obstacles to the advancement of

public health and sanitation. The state's failure to control infectious disease was affected by both external factors, including the limits of medical technology, a limited tax base from which to raise revenue, and India's disease ecology, and by the nature of being a colonial government, which privileged trade and fiscal responsibility to an imperial center over the health care of Indians.[39] Any large-scale public health campaign would have presupposed intensive financial, institutional, and educational commitments and resources. The Government of India scored poorly on all three. Whereas the well-known Italian anti-malarial campaign of the 1900s was national in scale, to account for mobile people and mosquitoes, and provided health education, access to care and medical personnel, and distribution of free quinine to the poor, the Government of India did none of these consistently, despite its admiration of Italy's efforts.[40] It was not that health officials in India did not recognize the prerequisites of effective policy; rather, disease control was subordinated to the broader economic and political concerns of the colonial state. F. P. Mackie, as Officiating Public Health Commissioner, stated: 'At the risk of reiteration I must emphasise that the *preliminary* steps of a successful public health policy are (1) health education of the population, (2) the arousing of a public health conscience and a conception of citizenship amongst the electorates and the local administrative authorities, and (3) the abolition of retrograde social customs.'[41] He argued that unless due regard was given to these basic principles in framing India's sanitary policy, 'isolated measures devised against particular diseases are bound to have only a partial success; they partake of the nature of short-cuts – they are longest in the end'.[42]

Medical research was one of the few beacons of light in the government's policies, but the disconnect between India's production of knowledge and its application continued to grow and became increasingly apparent to both international and local health experts. Public health officials in India often lamented India's negligence and inconsistency in the practical application of the very knowledge it happily displayed before international audiences. The treatment and research of disease were limited by diverse research objectives, colonial attitudes, lobby interests, and political and economic constraints.[43] Political and economic factors often determined the level of urgency a disease elicited, structured research priorities, and influenced the application of solutions identified through research.[44] The government hit a stumbling block when it came to the last.

The application of research remained limited, partly because of a lack of political will and partly because of colonial finances.[45] Public health officials were acutely aware of the impediment that the lack of health care financing posed. J. D. Graham believed that

> the State, both centrally and provincially, will require to allocate in the future larger sums to health than she has done in the past if real headway is to be made in India in the prevention of disease – essentially a State matter; – and further that, in the fulness [*sic*] of time, some considerable reorganisation of the state administration of public health and prevention of disease will be essential, if this Indian Empire is to come into line with similar world federations of States in such matters.[46]

Medical opinion in India held that the waste of life and its resultant consequences could be prevented at a cost small in comparison to the public health expenditure that would be saved by preventing disease.[47] However, the opinions of public health officials did not always determine policy at the center. While the changing political context after World War I created a greater demand for policies to reflect Indian rather than British interests, the interwar period saw a significant decline in state investment in the economy, including in rail building and agriculture, especially since there was a foreseeable British withdrawal from India.[48] If the colonial state was less inclined following the war and Devolution to invest in India's economy, it was even less interested in investing in India's health. Large-scale prevention probably would have required a fundamental change in the fiscal and budgetary structures of the colonial economy.[49]

India's health suffered from a lack of administrative capacity as well, meaning the extent to which the Government of India had the bureaucratic, fiscal, and statutory ability or power to implement precautions. Quarantine was less expensive and faster to implement than sanitary reform, making it the choice of poorer countries.[50] In a similar vein, it was easier logistically and administratively for the central government in India to focus on sanitary enclaves or on specific groups of people. Mackie argued that the natural history of five of the principal diseases of India – malaria, plague, cholera, smallpox, and kala-azar – had been sufficiently elucidated to enable their eradication 'if only money, energy and cooperation were combined in one grand attack' of prevention.[51] Bureaucratic inertia seeped into the government's malaria policies as well. Even as alternative forms of cinchona derivatives become more readily available and as their efficacious properties were better understood, the Government of India moved slowly in reviewing the new standards advocated by the League of Nations Health Organisation.

The priorities of imperial governance acted as another constraint on the options the Government of India deemed feasible.[52] With Devolution, the center retained the high-priority powers it wanted, such as finance and revenue, and devolved, or rather relegated, public health to the provincial and local levels. Devolution revealed the low priority that the colonial government had accorded India's health status by reducing the central government's public health jurisdiction. Another major shortcoming of the devolution policy was the lack of a coordinating central agency, which could ensure greater uniformity in the development of health policies across provinces.[53] Health officials believed that Devolution had particularly disadvantaged endemic disease control. In terms of international health relations, the imperatives of public health and economic diplomacy would keep India responsive to international opinion.[54] However, at the provincial level, legislatures had been able to act unfettered by central control, often creating a situation of uncoordinated and unequal action across provinces. India lacked coordinated control over the spread of epidemics across internal borders.[55] According to the British Social Hygiene Council, the provincialization of sanitation and public health in India had halted plans to create a unified public health policy for the whole of India.[56] The council noted significant divergences in policies of various provinces once the public health departments were made independent of central government control.

'Responsible government' meant that the Government of India could justify irresponsibility in terms of public health and sanitation, which after 1922 fell under the purview of the provinces. If risk governance concerns itself with the broader sociopolitical environments in which risk originates or is perceived and involves both citizens, or in this case colonial subjects, and political representatives, then it necessarily deals explicitly with political choices.[57] In supporting the devolution of public health to the provinces, the central government took a political position to transfer not only its ability to 'govern' risk but also its ability to manage risk by controlling health conditions and measures. The central government could not, thus, conduct effective disease control at an all-India level. Its public health responsibilities were limited by the 1919 Government of India Act to overseeing medical research and to the control of diseases to prevent their spread beyond India's borders. While the 1919 Act may have been politically expedient, it hindered the advancement of public health in India by preventing the formation of a strong central public health organization and by eliminating the ability of the central health officials to compel provinces to undertake certain measures in the interest of health and disease control.

Underneath these tangible obstacles to policy development lay a more insidious one. Colonial attitudes toward Indians, which were lodged within larger ideologies of empire, prevented a more concerted effort to address the determinants of disease. Public health policy had resulted from eighteenth- and nineteenth-century practices of empire with all of their inherent contradictions. Liberal imperialism embodied on the one hand a commitment to liberty, an ideology of toleration, and a belief in national self-determination. On the other, it adopted paternalism as a form of governance, unsympathetic views of diversity, assumptions of inequality of power, and a refusal to see India as a nation and Indians as a national people. These contradictions were reconciled by blaming Indians for their sanitary incorrigibility.[58] Britain had two simultaneous strategies for justifying rule in India – one stressing similarities between Indians and Britons and the other, which predominated from 1858 to 1918, emphasizing difference (of race, history, gender, society).[59] It was this ideology of difference, which pervaded official public health discourse. Policy decisions resulted from 'tensions of empire' particular to European imperialism of the nineteenth and twentieth centuries. There was constant conflict between the universalizing claims of imperialist ideology and the actual nature of conquest, the resultant limitations placed on rulers because of an ideology of difference, and varying degrees of exploitation and domination in colonial endeavors.[60] A rule based on difference meant that the colonial state would never fulfill its normalizing, universalizing mission, whether in the social, political, economic, or health domain.[61]

The attitudinal limitations on policy also reflected tensions between the Orientalizing and universalizing aspects of Western medicine.[62] Western medicine, if applied properly, could prevent and treat the ills of Indians. Anti-cholera and anti-plague inoculations were viewed as efficacious in reducing mortality and morbidity; quinine could treat malaria and reduce its debilitating effects; and yellow fever inoculations could offer immunity to an unexposed population. But at the same time that health officials saw hope through the vehicles of medical technologies and drugs, they also viewed Indian customs, hygienic practices, and intelligence as limiting factors in the application of modern medical techniques. Yet colonial rulers in India did not address the social context of behavior, which had been critical to the acceptance of new medical principles and hygienic practices in England in the second half of the nineteenth century.[63] India's diseases and dirt became markers of difference at the same time that bacteriological explanations and discoveries in parasitology challenged this ideology with the notion that all bodies were potentially susceptible to germs and their vectors.[64] Assumptions about the 'sanitary conscience' of Indians, the 'dead-weight of centuries of inertia', and supposed stifling racial and caste customs continued to influence the public health establishment into World War II. Government legislation was futile without the necessary education and propaganda to convince Indians of the merits of certain public health or sanitary measures, yet the success of health education was hampered by officials' perceptions, real or imagined, of Indian customs and habits.[65]

Another fundamental problem lay in the exclusion of public health from the concept of politics. Health officials in India did believe that systems of governance and laws enacted had a close relationship with public health and that economic conditions were intimately connected to the health of India's population. However, colonial 'politics' espoused a narrowness of vision that saw politics' purview as questions of franchise, representation, and forms of government. As long as this view persisted and dominated government actions, progress in public health and sanitation would remain sporadic.[66] The Government of India as a whole needed to see the health conditions of Indians as integral to any conception of politics. Only then could the government implement sustained campaigns to prevent and treat malnutrition and preventable diseases.

Given this multiplicity of obstacles, health policies to prevent and control disease continued to focus on the short-term. Experts knew that lack of sufficient food and drought exacerbated disease and that until the economic standard of Indians was raised, government and society could not progress in solving the social and health evils of India in any lasting way.[67] Government administration of inoculations and quinine was cheaper and logistically simpler than addressing the underlying sanitary and social causes of disease. In 1937, out of 23.5 million people living in urban areas in British India, only 12.7 million lived in towns with a protected water supply. Rural areas fared even worse, procuring water from tanks, reservoirs, rivers, and canals where people often bathed and washed soiled clothes. These water resources were liable to pollution and carried cholera infection to people living in banks along the rivers.[68] Sanitary reform in India was relegated to a secondary position behind central policies focusing on prophylactics and bacteriologically driven medicine.[69] Indeed, the emphasis on tropical medicine that emanated from the metropolitan center, and which was reinforced by the training of colonial medical officers in London and Liverpool, diverted resources and efforts away from dealing with the socio-economic determinants of disease and toward technological fixes and curative treatments.[70] The need to improve the overall health status of Indians rarely factored into policy discussions on a strategic level. Instead, state perceptions of risk created the need to make policies to prevent and control infectious disease, while domestic and international politics, the protection of trade, and epidemiological security structured the content of infectious disease policy in British India.

# Epilogue: Swine Flu Redux

It is not necessary to reach very far back in history to understand that government perceptions of risk profoundly influence the timing and content of disease control policies. Just a few years ago, in March of 2009, unusual cases of influenza-like illness began appearing in worrisome numbers in Mexico. Later identified as H1N1, this novel strain of influenza was quickly branded the 'swine flu'. This undesirable moniker led people to believe that eating pork could spread the disease to humans, causing the price of hogs to fall in the Chicago Mercantile Exchange. Outcries from the American pork industry ensued as countries began imposing trade restrictions on the American product. American hog farmers accused foreign governments of exploiting the outbreak to provide cover for their farmers in the midst of a global economic crisis. US trade authorities needed to convince countries to reverse the restrictions quickly, or bad policy would become permanent. Ron Kirk, the United States Trade Representative, urged America's trading partners to make decisions based on science and international trade obligations to avoid unnecessary trade disruptions.[1] Even into the fall (and start of flu season), fears of eating pork persisted. On 21 October 2009, Senator Claire McCaskill of Missouri asked Kathleen Sebelius, Secretary of Health and Human Services, to state for the record that eating pork did not cause H1N1.[2] Senator McCaskill worried about not only the epidemiological risk of H1N1 spread but also the economic impact of this disease on her pork producer constituents. As had happened with epidemics that had come before, initial responses to the H1N1 outbreaks were strongly influenced by trade jockeying and domestic politicking.

H1N1 was not immune to the age-old trifecta of travel restrictions, trade embargoes, and quarantine. In the spring of 2009, the number of new cases of H1N1 in Mexico continued rising at an alarming

158

rate, provoking a severe slowdown in air traffic to and from Mexico. President Obama tried to stem the panic, emphasizing that H1N1 spread was 'obviously a cause for concern . . . but it is not a cause for alarm'.[3] Other countries had different ideas. Argentina, Cuba, and others suspended flights to and from Mexico, while major tour operators in Britain, Canada, Germany, and France also canceled flights. The European Union Health Commissioner, Androulla Vassiliou, advised Europeans to delay non-essential travel to affected parts of Mexico and the US. In April, around 2000 flights per day were being canceled, partly due to certain countries' flight restrictions and partly because of travelers canceling their plans in the hope of avoiding contamination, either on board or while in Mexico.

Canceling flights, however, does little to prevent the spread of an influenza pandemic. Even if air travel were to be reduced in major cities by 99.9 percent immediately following the first case of influenza, this would postpone a pandemic by several weeks but, otherwise, do little to halt its spread. Scientific models suggest that it is almost impossible for countries to contain a novel strain of influenza.[4] Yet models will not prevent countries, which fear potential disaster, from employing whatever countermeasures are at their disposal to protect their populations, economies, and infrastructure.

The market, like governments, is often more sensitive to human emotion than to scientific data. Pork producers fared no better than the travel industry as China, Russia, and Ukraine instituted bans on pork and pork products from Mexico and from the three US states where swine flu cases had appeared. Azerbaijan went one step further and prohibited all North American livestock products. By late April, Egypt had begun culling pigs purportedly to prevent the disease, despite the World Health Organization's statement that there was no scientific evidence indicating that pigs could infect people. Egypt's policy to kill 300,000 pigs was condemned as another example of sectarian discrimination, with the Egyptian government using swine flu as an opportunity to hurt Egyptian pig farmers economically, most of whom are Coptic Christians.[5]

As countries and people became increasingly alarmed at the rate and extent of spread of H1N1, governments turned to quarantine in an effort to stave off infection of their populations. Health and transportation authorities in Indonesia, South Korea, Singapore, Thailand, Japan, and the Philippines brought out their arsenal of thermal scanners to screen passengers arriving from North America for signs of fever.[6] In Australia, shortly after several crew members of a cruise ship exhibited symptoms of swine flu, the health authorities in New South Wales announced that all cruise ships in Australian waters would be treated as 'infected' and appropriate measures would be taken, including not allowing passengers to disembark until the ship was cleared as 'uninfected'.[7] In May, China confined 21 students and 3 teachers from Maryland in their hotel rooms in Kaili, China, because they had sat within four rows of a fellow airplane passenger *suspected* of having swine flu. Consequently, the US State Department issued a travel alert to Americans in July and again in September about the possibility of quarantine in China if travelers were suspected of harboring the illness. The State Department received thousands of reports between May and August of American citizens quarantined in China in less than ideal conditions. In some instances, children were separated from parents, minors were administered medicines without parental consent, telephones were unavailable, and sanitary drinking water and food were not to be had. While other countries, such as Japan, had realized the futility of such draconian measures, China seemed to be stepping up these precautions without regard to their effects on tourism or business. In May, 240 guests and 100 employees at the Metropark Hotel in Hong Kong were quarantined for seven days and could not leave the hotel. If China had erred in the wrong direction in its delayed reporting of SARS in 2002, the government seemed to swing to the other extreme, overcompensating for prior accusations of public health negligence.[8]

Even when epidemiological risk seems to be the foremost consideration in formulating epidemic disease policies, other factors can be equally, if not more, important, as illustrated by the US' decision in 2009 not to use adjuvants in its H1N1 vaccine supply. Adjuvants are vaccine additives that confer two important benefits: they elicit a stronger immune response, making the vaccine more effective, and they require significantly less vaccine stock per vaccine, which increases the number of people limited supplies of the vaccine can reach. A study by Purdue University issued in October 2009 predicted that 59–69 percent of the population in America would be infected with H1N1 and that 40 percent of those infected would become ill.[9] Given the significant shortage of vaccine in the fall of 2009, just as the US epidemic was reaching its second peak in October, the decision not to use adjuvants, thereby effectively reducing the domestic vaccine supply, becomes even more controversial in retrospect.

The Department of Health and Human Services (HHS) had not *a priori* decided against the use of adjuvants. In May and July of 2009, HHS had ordered close to $700 million of adjuvant in the initial development stages of the H1N1 vaccine to determine the safety of adjuvants and their effects on the immune system's response.[10] At some point before vaccine production began, the decision was made not to use adjuvants for vaccines administered in the US, even though adjuvanted H1N1 vaccines were being widely used in Europe, the Middle East, Asia, and Canada. The WHO emphasized the positive safety data on the adjuvanted version and encouraged countries to use the adjuvanted vaccine to increase the global supply. According to the WHO, H1N1 vaccine adjuvants would double the yield for the same quantity of antigen. One of the vaccine adjuvants being used in the H1N1 vaccine had been a component of seasonal influenza vaccines in Europe for years and had shown no adverse effects.[11] The US was seemingly hesitant to risk a domestic backlash against adjuvants, particularly given the increasingly vocal and growing numbers of vaccine-hesitant parents and the still-present memories of the 1976 swine flu affair, which continue to haunt the US vaccine establishment today.[12]

As the adjuvant decision shows, epidemiology is by no means the only or primary factor running through policymakers' minds during times of epidemiological crisis. Fears of disease spread can damage industries, hurt productivity of workers, and put at risk a country's critical infrastructure. They can place policymakers under great strain to respond quickly, even if that response is inadequate or inappropriate. While the relationship between state and society, the availability of resources, the strength of political and civil institutions, levels of economic development, and administrative capacity all influence how states respond to epidemics, these factors are insufficient to explain differences in responses among countries dealing with epidemic crises.[13] Even with the new International Health Regulations in place, national responses evidenced considerable variation in adherence to those regulations. The 2005 International Health Regulations had been hailed as representing a paradigm shift in international disease control by changing the focus from controlling disease spread at points of entry (ports, airports, borders) to reducing disease incidence within borders through surveillance and intervention. Yet the initial responses to H1N1 around the world underscore the limitations of regulations not ratified by legislatures of signatory countries and reveal the ingrained patterns of reaction that states have.[14]

The spectrum of responses to H1N1 suggests that perceptions of and reactions to the varied risks associated with epidemic disease outbreaks continue to be critical factors in shaping disease prevention and preparedness and response policies today. Risk, in its myriad forms, is central to infectious disease policy decisions. Governments, with access to a similar body of scientific information, have diverse perceptions of risk, different assumptions about what that risk entails, differing levels of risk tolerance, and disparate ideas on how to minimize risk. Our current approaches to epidemic disease control continue to be influenced by the practices of the nineteenth and twentieth centuries and by government perceptions of risk which, intertwined with public fears, continue to buttress unexamined assumptions that pervade the policy-making process. Until governments develop a common framework that thoroughly considers different types of risks as factors influencing epidemic disease policy options, it will remain difficult to mitigate against the sometimes extreme responses perceptions of risk provoke.

# Notes

## Introduction

1. Ulrich Beck, *World Risk Society* (Cambridge: Polity Press, 1999), p. 135.
2. Allan M. Brandt, 'Behavior, Disease, and Health in the Twentieth-Century United States: The Moral Valence of Individual Risk', *Morality and Health*, Allan M. Brandt and Paul Rozin (eds) (New York and London: Routledge, 1997), p. 54.
3. Kofi Annan, 'Secretary-General Proposes Global Fund for Fight Against HIV/AIDS and Other Infectious Diseases at African Leaders Summit', http://www.theglobalfund.org/en/history (accessed 8 December 2008).
4. Jeffery Sachs, 'Macroeconomics and Health: Investing in Health for Economic Development' (Geneva: World Health Organization, 2001), p. 16.
5. William R. Rodriguez, 'Interview with Robert C. Bollinger and William R. Rodriguez', http://www.globalur.com (accessed 8 December 2008).
6. Jeffery Sachs, 16 April 2002, http://www.wilsoncenter.org (accessed 8 December 2008).
7. National Intelligence Council, 'The Global Infectious Disease Threat and Its Implications for the United States', January 2000. National Intelligence Council, 'The Next Wave of HIV/AIDS: Nigeria, Ethiopia, Russia, India, and China', September 2002.
8. 'PEPFAR Funding', http://www.pepfar.gov (accessed 2 January 2012).
9. Julio Frenk, 'Economic Crises and Health: Risk or Opportunity'? Lecture given at Center for History and Economics, Harvard University, Cambridge, MA, 2 December 2008.
10. Michel Foucault, *Security, Territory, Population: Lectures at the Collège de France, 1977–1978*, Michel Senellart (ed.) (New York: Picador, 2004), p. 6.
11. Cf. David P. Fidler, 'Towards a Global *ius pestilentiae*: The Functions of Law in Global Biosecurity', *Global Biosecurity: Threats and Responses*, Peter Katona, John P. Sullivan and Michael D. Intriligator (eds) (London: Routledge, 2010).
12. *Proceedings of the Government of India* (hereafter GOI), Home (Sanitary), January 1901, IOR/p/6118. (For all government proceedings, the month cited refers to the month under which the proceeding was included. If the year not divided into monthly entries, then the month cited indicates the date of the proceeding. IOR refers to the India Office Records at the British Library.)
13. Cf. Charles E. Rosenberg, *The Cholera Years: The United States in 1832, 1849, and 1866* (Chicago: The University of Chicago Press, 1987 [1962] and Charles E. Rosenberg, 'Cholera in 19th-Century Europe: A Tool for Social and Economic Analysis', *Explaining Epidemics* (Cambridge: Cambridge University Press, 1992).
14. Rosemary A. Stevens, 'Introduction', *History and Health Policy in the United States: Putting the Past Back In*, Rosemary A. Stevens, Charles E. Rosenberg, and Lawton R. Burns (eds) (New Brunswick: Rutgers University Press, 2006), p. 3.

15. Charles E. Rosenberg, 'Anticipated Consequences: Historians, History, and Health Policy', *History and Health Policy in the United States: Putting the Past Back In*, Rosemary A. Stevens, Charles E. Rosenberg, and Lawton R. Burns (eds) (New Brunswick: Rutgers University Press, 2006), pp. 13–14. Allan M. Brandt, 'From Analysis to Advocacy: Crossing Boundaries as a Historian of Health Policy', *Locating Medical History: The Stories and Their Meanings*, Frank Huisman and John Harley Warner (eds) (Baltimore: The Johns Hopkins University Press, 2004), p. 463.
16. Roy MacLeod, 'Introduction', *Disease, Medicine, and Empire: Perspectives on Western Medicine and the Experience of European Expansion*, Roy MacLeod and Milton Lewis (eds) (London: Routledge, 1988). Cf. Mark Harrison, *Public Health in British India: Anglo-Indian Preventive Medicine, 1859–1914* (Cambridge: Cambridge University Press, 1994).
17. Harrison, *Public Health in British India*, p. 137.
18. Sunil S. Amrith, *Decolonizing International Health: India and Southeast Asia, 1930–65* (Houndmills, Basingstoke: Palgrave Macmillan, 2006), pp. 4, 11.
19. Cf. Biswamoy Pati and Mark Harrison, 'Introduction', Health, Medicine and Empire: Perspectives on Colonial India, Biswamoy Pati and Mark Harrison (eds) (New Delhi: Orient Longman Limited, 2001). Daniel R. Headrick, *The Tools of Empire: Technology and Imperialism in the Nineteenth Century* (New York, Oxford University Press, 1981). Jane Buckingham, *Leprosy in Colonial South India: Medicine and Confinement* (New York: Palgrave Macmillan, 2002). Gyan Prakash, *Another Reason: Science and the Imagination of Modern India* (Princeton: Princeton University Press, 1999). Maneesha Lal, '"The Ignorance of Women is the House of Illness": Gender, Nationalism, and Health Reform in Colonial North Indian', *Medicine and Colonial Identity*, Mary P. Sutphen and Bridie Andrews (eds) (London: Routledge, 2003). Alison Bashford, *Imperial Hygiene: A Critical History of Colonialism, Nationalism and Public Health* (New York: Palgrave Macmillan, 2004). David Arnold, *Science, Technology and Medicine in Colonial India* (Cambridge: Cambridge University Press, 2000). Alison Bashford and Claire Hooker (eds), *Contagion: Epidemics, History and Culture From Smallpox to Anthrax* (Annandale: Pluto Press Australia, 2002). Soma Hewa, *Colonialism, Tropical Disease and Imperial Medicine: Rockefeller Philanthropy in Sri Lanka* (Lanham: University Press of America, 1995). Poonam Bala (ed.), *Biomedicine as a Contested Site: Some Revelations in Imperial Contexts* (Lanham: Lexington Books, 2009). Ishita Pande, *Medicine, Race and Liberalism in British Bengal: Symptoms of Empire* (London: Routledge, 2010).
20. MacLeod, pp. 10–11.
21. Cf. Headrick and David Washbrook, 'Orients and Occidents: Colonial Discourse Theory and the Historiography of the British Empire', *The Oxford History of the British Empire: Volume V: Historiography*, Robin W. Winks (ed.) (Oxford: Oxford University Press, 1999).
22. Ann Laura Stoler and Frederick Cooper, 'Between Metropole and Colony: Rethinking a Research Agenda' *Tensions of Empire: Colonial Cultures in a Bourgeois World*, Frederick Cooper and Ann Laura Stoler (eds) (Berkeley: University of California Press, 1997), pp. 3–4, 20.
23. David Armitage, 'Three Concepts of Atlantic History', *The British Atlantic World, 1500–1800*, David Armitage and Michael J. Braddick (eds) (Houndmills, Basingstoke: Palgrave Macmillan, 2002), pp. 15–16.

24. Sugata Bose, 'Space and Time on the Indian Ocean Rim: Theory and History', *Modernity and Culture: From the Mediterranean to the Indian Ocean*, Leila Tarazi Fawaz, C. A. Bayly, and Robert Ilbert (eds) (Columbia University Press, 2002), p. 369. Cf. Sugata Bose, *A Hundred Horizons: the Indian Ocean in the Age of Global Empire* (Cambridge: Harvard University Press, 2006).
25. Akira Iriye, *Cultural Internationalism and World Order* (Baltimore: The Johns Hopkins University Press, 1997), pp. 177–8.
26. C. A. Bayly, '"Archaic" and "Modern" Globalization in the Eurasian and African Arena, c. 1750–1850', *Globalization in World History*, A. G. Hopkins (ed.) (New York: W.W. Norton, 2002), p. 63. Cf. Philip Curtin, *The World and the West: The European Challenge and the Overseas Response in the Age of Empire* (Cambridge: Cambridge University Press, 2000).
27. Cf. K. N. Chaudhuri, *Trade and Civilisation in the Indian Ocean: An Economic History from the Rise of Islam to 1750* (Cambridge: Cambridge University Press, 1985). Bernard Bailyn, *Atlantic History: Concept and Contours* (Cambridge: Harvard University Press, 2005).
28. Peter Bennett', Understanding responses to risk: some basic findings', *Risk Communication and Public Health*, Peter Bennett and Sir Kenneth Calman (eds) (Oxford: Oxford University Press, 1999), p. 14.
29. Kathleen J. Tierney, 'Toward a Critical Sociology of Risk', *Sociological Forum*, 14.2 (1999), p. 217. John J. Cohrssen and Vincent T. Covello, *Risk Analysis: A Guide to Principles and Methods for Analyzing Health and Environmental Risks* (Springfield, VA: United States Council on Environmental Quality, 1989), pp. 1–2. Paul C. Stern and Harvey V. Fineberg (eds), *Understanding Risk: Informing Decisions in a Democratic Society* (Washington, DC: National Academy Press, 1996), pp. 2–3, 27.
30. Cohrssen and Covello, p. 2.
31. Beck, pp. 136–7.
32. Nick Pidgeon, 'Risk, Uncertainty and Social Controversy: From Risk Perception and Communication to Public Engagement', *Uncertainty and Risk: Multidisciplinary Perspectives*, Gabriele Bammer and Michael Smithson (eds) (London: Earthscan, 2008), p. 359. Vincent T. Covello and Branden B. Johnson, 'The Social and Cultural Construction of Risk: Issues, Methods, and Case Studies', *The Social and Cultural Construction of Risk: Essays on Risk Selection and Perception*, Branden B. Johnson and Vincent T. Covello (eds) (Dordrecht: D. Reidel Publishing Company, 1987), pp. vii–viii.
33. Mary Douglas and Aaron Wildavsky, *Risk and Culture: An Essay on the Selection of Technical and Environmental Dangers* (Berkeley: University of California Press, 1982), p. 14.
34. Edward O. Laumann and David Knoke, *The Organizational State: Social Choice in National Policy Domains* (Madison: The University of Wisconsin Press, 1987), p. 39.
35. Lee Clarke and James F. Short, Jr. 'Social Organization and Risk: Some Current Controversies', *Annual Review of Sociology*, 19 (1993), pp. 379, 385.
36. Tierney, pp. 217, 223, 236.
37. Anthony Giddens, *Modernity and Self-Identity: Self and Society in the Late Modern Age* (Stanford: Stanford University Press, 1991), pp. 4, 121.
38. Gabe Mythen, *Ulrich Beck: A Critical Introduction to the Risk Society* (London: Pluto Press, 2004), pp. 157–8, 180–2. Tierney, p. 216.

39. Deborah Lupton, *The Imperative of Health: Public Health and the Regulated Body* (London: Sage Publications, 1995), p. 77. Alan Petersen and Deborah Lupton, *The New Public Health: Health and Self in the Age of Risk* (St. Leonards, Australia: Allen and Unwin, 1996), p. xiii.

40. William Coleman, *Death is a Social Disease: Public Health and Political Economy in Early Industrial France* (Madison: The University of Wisconsin Press, 1982), pp. 34, 124. Cf. Constance A. Nathanson, *Disease Prevention as Social Change: the state, society, and public health in the United States, France, Great Britain, and Canada* (New York: Russell Sage Foundation, 2007).

41. John Eyler, *Victorian Social Medicine: The Ideas and Methods of William Farr* (Baltimore: The Johns Hopkins University Press, 1979), pp. 95–6.

42. Cf. Peter Baldwin, *Contagion and the State in Europe, 1830–1930* (Cambridge: Cambridge University Press, 1999). Nathanson, pp. 233–4, 240.

43. Nathanson, p. 233. Cf. Mariola Espinosa, *Epidemic Invasions: Yellow Fever and the Limits of Cuban Independence, 1878–1930* (Chicago: University of Chicago Press, 2009).

44. Andrew T. Price-Smith, 'Ghosts of Kigali: Infectious Disease and Global Stability at the Turn of the Century', *Plagues and Politics: Infectious Disease and International Policy*, Andrew T. Price-Smith (ed.) (Houndmills, Basingstoke: Palgrave Macmillan, 2001), p. 173. Robert L. Ostergard, Jr., 'HIV/AIDS, State Capacity, and the Threat to National and International Security: A Theoretical Overview', *HIV/AIDS and the Threat to National and International Security*, Robert L. Ostergard, Jr. (ed.) (Houndmills, Basingstoke: Palgrave Macmillan, 2007), pp. 66–7.

45. Price-Smith, 'Ghosts of Kigali: Infectious Disease and Global Stability at the Turn of the Century', p. 175.

46. Cf. Charles Rosenberg, 'Banishing Risk: Continuity and Change in the Moral Management of Disease', *Morality and Health*, Allan M. Brandt and Paul Rozin (eds) (New York and London: Routledge, 1997). Brandt, 'Behavior, Disease, and Health in the Twentieth-Century United States'. Charles E. Rosenberg, 'Pathologies of Progress: The Idea of Civilization as Risk', *Bulletin of the History of Medicine*, 72.4 (1998), pp. 714–30.

47. Brandt, 'Behavior, Disease, and Health in the Twentieth-Century United States', pp. 60, 62–3. Allan M. Brandt, *The Cigarette Century: The Rise Fall, and Deadly Persistence of the Product that Defined America* (New York: Basic Books, 2007), pp. 211–12. Allan M. Brandt, '"Just Say No": Risk, Behavior, and Disease in Twentieth-Century America', *Scientific Authority and Twentieth-Century America*, Ronald G. Walters (ed.) (Baltimore: Johns Hopkins University Press, 1997), pp. 82, 88.

48. Cf. Aileen J. Plant, 'When Action Can't Wait: Investigating Infectious Disease Outbreaks', *Uncertainty and Risk: Multidisciplinary Perspectives*, Gabriele Bammer and Michael Smithson (eds) (London: Earthscan, 2008).

49. Cf Andrew Lakoff and Stephen J. Collier (eds), *Biosecurity Interventions: Global Health and Security in Question* (New York: Columbia University Press, 2008). Cf. Alison Bashford (ed.), *Medicine at the Border: Disease, Globalization and Security, 1850 to the Present* (New York: Palgrave Macmillan, 2006).

50. Susan Peterson, 'Human security, National Security, and Epidemic Disease', *HIV/AIDS and the Threat to National and International Security*, Robert L. Ostergard, Jr. (ed.) (Houndmills, Basingstoke: Palgrave Macmillan, 2007),

pp. 39–40. Andrew T. Price-Smith, *Contagion and Chaos: Disease, Ecology, and National Security in the Era of Globalization* (Cambridge: MIT Press, 2009), p. 4.

51. Donald F. Thompson and Renata P. Louie, 'Cooperative Crisis Management and Avian Influenza: A Risk Assessment Guide for International Contagious Disease Prevention and Risk Mitigation', March 2006: 28–29, http://www.ndu.edu/ctnsp/Def_Tech (accessed 3 April 2009).

52. Ostergard, p. 74.

53. This study focuses on qualitative understandings of risk, since that was how most officials approached health security at the time. There were some attempts in the 1920s and 1930s to quantify the economic costs of poor health and disease.

54. David P. Fidler, 'The Globalization of Public Health: The First 100 Years of International Health Diplomacy', *World Health Organization*, 79 (2001), p. 847.

55. For discussion of natural rights and sovereignty, see Richard Tuck, *The Rights of War and Peace: Political Thought and the International Order from Grotius to Kant* (Oxford: Oxford University Press, 1999).

56. Gail Minault, *The Khilafat Movement: Religious Symbolism and Political Mobilization in India* (New York: Columbia University Press, 1982) Chapters 2 and 3. Ayesha Jalal, *Self and Sovereignty: Individual and Community in South Asian Islam Since 1850* (London: Routledge, 2000) 186, 188. Sumit Sarkar, *Modern India: 1885–1947* (London: Macmillan Press, 1989) 165. David Washbrook, 'The Rhetoric of Democracy and Development in Late Colonial India', *Nationalism, Democracy, and Development*, Sugata Bose and Ayesha Jalal (eds) (Delhi: Oxford University Press, 1997), p. 44.

57. Niklas Luhmann, *Risk: A Sociological Theory* (New Brunswick: Aldine Transaction, 2002), pp. 19–20.

58. Ibid., p. 23.

59. Ibid., p. 29.

60. Cf. Judith Walzer Leavitt, '"Typhoid Mary" Strikes Back Bacteriological Theory and Practice in Early Twentieth-Century Public Health', *Isis*, 83.4 (1992).

61. F. Norman White, *The Prevalence of Epidemic Disease and Port Health Organisation and Procedure in the Far East* (Geneva: League of Nations, 1923), p. 148.

62. John Farley, 'Parasites and the Germ Theory of Disease', *Framing Diseases: Studies in Cultural History*, Charles E. Rosenberg and Janet Golden (eds) (New Brunswick: Rutgers University Press, 1992), p. 42.

63. Frank G. Clemow, *The Geography of Disease* (Cambridge: University Press, 1903), pp. 520–1. In current usage, hyperendemic signifies the habitual presence of a disease at a high level of incidence at all ages.

64. Ira Klein, 'Plague, Policy and Popular Unrest in British India', *Modern Asian Studies* 22.4 (1988), p. 735.

65. Ira Klein, 'Death in India, 1871–1921', *Journal of Asian Studies*, 32 (1973), pp. 645–6.

66. 'The Need for a Public Health for India', *Indian Medical Gazette*, 62 (1927), p. 575 (hereafter *IMG*).

67. David Arnold, 'Introduction: Tropical Medicine before Manson', *Warm Climates and Western Medicine: The Emergence of Tropical Medicine, 1500–1900*, David Arnold (ed.) (Amsterdam: Rodopi B.V., 1996), p. 4.

68. Michael Worboys, 'Germs, Malaria and the Invention of Mansonian Tropical Medicine: From "Diseases in the Tropics" to "Tropical Diseases"', *Warm Climates and Western Medicine: The Emergence of Tropical Medicine, 1500–1900*, David Arnold (ed.) (Amsterdam: Rodopi B.V., 1996), p. 196.
69. Fidler, 'The Globalization of Public Health', p. 847. David P. Fidler, *International Law and Infectious Diseases* (Oxford: Clarendon Press, 1999), p. 52.
70. David Arnold, 'Touching the Body: Perspectives on the Indian Plague, 1896–1900', *Selected Subaltern Studies*, Ranajit Guha and Gayatri Chakravorty Spivak (eds) (Oxford: Oxford University Press, 1988), p. 394.
71. George Rosen, *A History of Public Health* (New York: MD Publications, 1958), pp. 110–11. John Eyler, *Sir Arthur Newsholme and State Medicine, 1885–1935* (Cambridge: Cambridge University Press, 1997). Christopher Hamlin, *Public Health and Social Justice in the Age of Chadwick: Britain, 1800–1854* (Cambridge: Cambridge University Press, 1998), p. 335. Dorothy Porter, *Health, Civilization and the State: A History of Public Health from Ancient to Modern Times* (London: Routledge, 1999).
72. Partha Chatterjee, *Nationalist Thought and the Colonial World: A Derivative Discourse* (Minneapolis: University of Minnesota Press, 1986), p. 74.
73. Porter, *Health, Civilization and the State*, pp. 57–61. Sudipta Sen, *Distant Sovereignty: National Imperialism and the Origins of British India* (New York: Routledge, 2002) 87.
74. Partha Chatterjee, *The Nation and Its Fragments: Colonial and Postcolonial Histories* (Princeton: Princeton University Press, 1993), pp. 17–18. Uday Singh Mehta, *Liberalism and Empire: A Study in Nineteenth-Century British Liberal Thought* (Chicago: The University of Chicago Press, 1999), pp. 21, 190–9.
75. 'Western' medicine is used as a short-hand for allopathic (as opposed to homeopathic), non-indigenous medicine, while recognizing that systems of medicine in India and in Europe are neither wholly Eastern nor wholly Western, but rather a result of evolutions influenced by centuries-old exchanges of medical practices and texts among Europe, the Middle East, and Asia.
76. Richard Drayton, 'Science, Medicine, and the British Empire', *The Oxford History of the British Empire: Volume V: Historiography*, Robin W. Winks (ed.) (Oxford: Oxford University Press, 1999), pp. 271–4. Diana Wylie, 'Disease, Diet, and Gender: Late 20th century Perspectives on Empire', *The Oxford History of the British Empire: Volume V: Historiography*, Robin W. Winks (ed.) (Oxford: Oxford University Press, 1999), pp. 279–80.
77. In the nineteenth century, public health measures were not well-funded or widely accessible in Europe either. Roger Jeffery, *The Politics of Health in India* (Berkeley: University of California Press, 1988), pp. 19–20.
78. Sandeep Sinha, *Public Health Policy and the Indian Public: Bengal 1850–1920* (Calcutta: Vision Publications, 1998), pp. 42, 89.
79. David Arnold, *Colonizing the Body: State Medicine and Epidemic Disease in Nineteenth-Century India* (Berkeley: University of California Press, 1993).
80. Prakash Chapter 5. Cf. Michel Foucault, *The Birth of Biopolitics: Lectures at the Collège de France, 1978–1979*, Michel Senellart (ed.) (New York: Picador, 2004). Michel Foucault, *Security, Territory, Population: Lectures at the Collège de France, 1977–1978*, Michel Senellart (ed.) (New York: Picador, 2004).

81. Bernard Porter, *The Lion's Share: A Short History of British Imperialism, 1850–2004* (Harlow: Pearson, 2004), Chapter 4. Susan Kingsley Kent, *Gender and Power in Britain, 1640–1990* (London: Routledge, 1999), pp. 236–9.
82. Christopher Hamlin, 'State Medicine in Great Britain', *The History of Public Health and the Modern State*, Dorothy Porter (ed.) (Amsterdam: Rodopi B.V., 1994).
83. Porter, *The Lion's Share*, Chapter 4. Kent, pp. 236–9.
84. Cf. Dipesh Chakrabarty, *Provincializing Europe: Postcolonial Thought and Historical Difference* (Princeton: Princeton University Press, 2000).
85. Sanchari Dutta, 'Plague, Quarantine and Empire: British-Indian Sanitary Strategies in Central Asia, 1897–1907', *The Social History of Health and Medicine in Colonial India*, Biswamoy Pati and Mark Harrison (eds) (London: Routledge, 2009), p. 86.
86. India's financial and military participation in World War I had earned India a seat in peace negotiations and the right to be a signatory to the treaty.
87. Cf. C. A. Bayly and Leila Fawaz, 'Introduction', *Modernity and Culture: From the Mediterranean to the Indian Ocean*, Leila Tarazi Fawaz, C. A. Bayly, and Robert Ilbert (eds) (Columbia University Press, 2002), pp. 14–16.
88. Rosenberg, 'Anticipated Consequences', pp. 13–14.
89. Imperial is used to refer to the British Empire as a whole and colonial to the Government of India and provincial governments in India.
90. The OIHP was established as an epidemiological information clearinghouse for states and public health administrations. It primarily occupied itself with plague, cholera, yellow fever, typhus, and smallpox. G. Abt, *Vingt-Cinq Ans d'Activité de L'Office International D'Hygiène Publique, 1909–1933* (Paris: Office International D'Hygiène Publique, 1933), p. 73. Nongovernmental organizations such as the Rockefeller Foundation and the International Red Cross Society also undertook public health works in India, but they did not significantly concern themselves with plague or cholera and worked on malaria only in specific localities. E. W. C. Bradfield, *An Indian Medical Review* (New Delhi: Government of India Press, 1938), pp. 258–61.
91. Judith M. Brown, 'Imperial Facade: Some Constraints upon and Contradictions in the British Position in India, 1919–35', *Transactions of the Royal Historical Society* series 5, 26 (1976), p. 37. The capital of India was transferred from Calcutta to New Delhi in 1911. Sugata Bose and Ayesha Jalal, *Modern South Asia: History, Culture, Political Economy* (New York: Routledge, 1998), pp. 102–3.
92. Stoler and Cooper.
93. Ronald Hyam, *Britain's Imperial Century, 1815–1914: A Study of Empire and Expansion* (Lanham: Barnes and Nobles Books, 1993).
94. Chatterjee, *Nationalist Thought and the Colonial World*, p. 145. Chatterjee, *The Nation and Its Fragments: Colonial and Postcolonial Histories*, p. 10.
95. Bose and Jalal (1998), pp. 129–31. Washbrook, 'The Rhetoric of Democracy and Development in Late Colonial India', pp. 41–3.
96. Washbrook, 'The Rhetoric of Democracy and Development in Late Colonial India', pp. 42–3.
97. Presidency was a term used to signify a political subunit within British India. Presidency and province are used interchangeably.
98. Rosen, pp. 332–5.

99. Part of the difficulty in recovering Indian influence on policy, particularly the impact of the lower classes, results from biases inherent in government archival documents and medical journals, which were written by civilian, military, and medical officials. They perceived Indian reaction to policies in terms of two extremes – either acquiescence or resistance. Popular sources, such as newspapers, journals, and pamphlets, tended to reflect the opinions of the middle- and upper-class Indian elite. Lauren Minsky, 'Pursuing Protection from Disease: The Making of Smallpox Prophylactic Practice in Colonial Punjab', *Bulletin of the History of Medicine*, 83.1 (2009), pp. 165–6.

100. Saurabh Mishra, *Pilgrimage, Politics, and Pestilence: The Haj from the Indian Subcontinent, 1860–1920* (New Delhi: Oxford University Press, 2011), pp. 20, 23.

101. Roger Jeffery, 'Doctors and Congress: The Roles of Medical Men and Medical Politics in Indian Nationalism', *The Indian National Congress and the Political Economy of India 1885–1985*, Mick Shepperdson and Colin Simmons (eds) (Aldershot: Avebury, 1988), p. 161. David Arnold, 'The 'Discovery' of Malnutrition and Diet in Colonial India' *The Indian Economic and Social History Review*, 31.1 (1994), pp. 24–6.

102. 'Report of the National Planning Committee, 1938' (New Delhi: Indian Institute of Applied Political Research, 1988 [1938]) 4, pp. 218–20.

103. Margaret Jones, *Health Policy in Britain's Model Colony: Ceylon (1900–1948)* (New Delhi: Orient Longman, 2004), p. 269. Mark Harrison, 'Public Health and Medical Research in India, c. 1860–1914', diss, Oxford University, 1991, p. 188.

104. Leprosy and smallpox are not considered here because they were viewed as primarily municipal and local matters, and it was recognized in the 1890s that the possibility of transmitting leprosy was small.

105. Warwick Anderson, 'Postcolonial Histories of Medicine', *Locating Medical History: The Stories and Their Meanings*, Frank Huisman and John Harley Warner (eds) (Baltimore: The Johns Hopkins University Press, 2004), p. 290.

106. Deepak Kumar, 'Perceptions of Public Health: A Study in British India', *Maladies, Preventives and Curatives: Debates in Public Health in India*, Amiya Kumar Bagchi and Krishna Soman (eds) (Kolkata: Tulika Books, 2005), p. 53. Achintya Kumar Dutta, '*Kala-Azar* in Assam: British Medical Intervention and People's Response', *Maladies, Preventives and Curatives: Debates in Public Health in India*, Amiya Kumar Bagchi and Krishna Soman (eds) (Kolkata: Tulika Books, 2005), p. 28. Toward the end of the war, the GOI and the Indian National Congress did begin planning for postwar measures to improve the health of Indians. Sunil Amrith, 'Rockefeller Foundation and Postwar Public Health in India', http://archive.rockefeller.edu/publications/ resrep/amrith.pdf (accessed 14 May 2007). Arabinda Samanta, Malarial Fever in Colonial Bengal, 1820–1939: Social History of an Epidemic (Kolkata: Firma KLM, 2002), p. 6.

107. Amrith, *Decolonizing International Health*, p. 47.

108. B. R. Tomlinson, *Political Economy of the Raj, 1914–1947: The Economics of Decolonization in India* (London: Macmillan Press, 1979), p. 56. Dietmar Rothermund, *India in the Great Depression, 1929–1939* (New Delhi: Manohar, 1992), p. 272.

# 1 All Eyes on India

1. David Arnold, 'The Indian Ocean as a Disease Zone, 1500–1950', *South Asia* (1991): 1–21.
2. C. A. Bayly, *The Birth of the Modern World, 1780–1914: Global Connections and Comparisons* (Malden: Blackwell Publishing, 2004), p. 234.
3. Arnold, 'The Indian Ocean as a Disease Zone', pp. 8–10.
4. Not all health officials in India agreed with European assessments of India being the source of cholera and plague. W. J. Simpson, 'Maritime Quarantine and Sanitation in Relation to Cholera', *The Practitioner: A Journal of Therapeutics and Public Health*, 48 (1892): 148–60. Paper prepared for the International Hygienic Congress held in London in August 1891.
5. Quarantine, which originally had entailed 40 days' detention, acquired a new meaning in the nineteenth century. The practice moved away from long, strict detentions and came to encompass the designation of sanitary surveillance and inspection stations, detention, and disinfection of vessels, people, baggage, and merchandise. Quarantine could be imposed on land or sea. Joseph Holt, *An Epitomized Review of the Principles and Practice of Maritime Sanitation* (New Orleans: L. Graham & Son, 1892), pp. 6–7, 11. When the term 'quarantine' is used alone, it refers to sea quarantine.
6. David P. Fidler, *International Law and Infectious Diseases* (Oxford: Clarendon Press, 1999) 115. Peter Baldwin, *Contagion and the State in Europe, 1830–1930* (Cambridge: Cambridge University Press, 1999) 7.
7. Bayly, *The Birth of the Modern World*, pp. 238, 243.
8. Fidler, *International Law and Infectious Diseases*, 9. W. F. Bynum, 'Policing Hearts of Darkness: Aspects of the International Sanitary Conferences', *History and Philosophy of the Life Sciences* 15 (1993): 426. I use Europe to refer to Continental Europe.
9. Ulrich Beck, *World Risk Society* (Cambridge: Polity Press, 1999), pp. 8, 15, 50, 142.
10. David P. Fidler, 'Public Health and International Law: the Impact of Infectious Diseases on the Formation of International Legal Regimes, 1800–2000', *Plagues and Politics: Infectious Disease and International Policy*, ed. Andrew T. Price-Smith (Houndmills, Basingstoke: Palgrave Macmillan, 2001), pp. 267–9. Mark Harrison, 'Disease, Diplomacy and International Commerce: The Origins of International Sanitary Regulation in the Nineteenth Century', *Journal of Global History*, 1 (2006): 197–8, 209.
11. Bynum, 'Policing Hearts of Darkness', p. 15.
12. Eleven European states and the Ottoman Porte attended the first conference in Paris. The General Treaty of Paris, signed in 1856, admitted the Sublime Porte, albeit with lesser status and authority, to 'participation in the advantages of the Public Law of Europe and the System of Concert attached to it'. Travers Twiss, *The Law of Nations considered as Independent Political Communities* (Oxford: University Press, 1861 [reprinted by Gaunt, 2000]).
13. Mark W. Zacher and Tania J. Keefe, *The Politics of Global Health Governance: United by Contagion* (Houndmills, Basingstoke: Palgrave Macmillan, 2008) 18.
14. Cf. Akira Iriye, *Cultural Internationalism and World Order* (Baltimore: The Johns Hopkins University Press, 1997). Unlike cultural internationalism, early public health internationalism did not aim to promote world peace.

15. Patrick Zylberman, 'Civilizing the State: Borders, Weak States and International Health in Modern Europe', *Medicine at the Border: Disease, Globalization and Security, 1850 to the Present*, ed. Alison Bashford (New York: Palgrave Macmillan, 2006), pp. 24, 34.

16. Baldwin, p. 126.

17. Britain was supported in its opposition to quarantine by most northern European countries as well as Italy. Norman Howard-Jones, *International Public Health between the Two World Wars: The Organizational Problems* (Geneva: World Health Organization, 1978), pp. 22, 40.

18. According to the 1897 Venice Convention, only if plague presented in epidemic form was the area declared infected. Mark Harrison, 'Public Health and Medical Research in India, c. 1860–1914', diss, Oxford University, 1991, 215.

19. Britain was averse to following disease control measures and even, at times, to participating in the conferences, when it potentially conflicted with British interests. Norman Howard-Jones, *The Scientific Background of the International Sanitary Conferences, 1851–1938* (Geneva: World Health Organization, 1975), p. 62. India sent its own delegates to some conferences, while at others, British delegates represented India's point of view. At the 1892 and 1894 conferences, delegates from both Britain and India were both present and had separate votes.

20. Howard-Jones, *The Scientific Background of the International Sanitary Conferences*, p. 9. Andrew Cunningham, 'Transforming Plague: The Laboratory and the Identity of Infectious Disease', *The Laboratory Revolution in Medicine*, Andrew Cunningham and Perry Williams (eds) (Cambridge: Cambridge University Press, 1992), pp. 210, 216–17. W. F. Bynum, *Science and the Practice of Medicine in the Nineteenth Century* (Cambridge: Cambridge University Press, 1994), p. 226. Despite the increasing bacteriological discoveries of the 1880s and 1890s, there was no simple shift from a physiological to ontological view of all diseases. Michael Worboys, *Spreading Germs: Disease Theories and Medical Practice, 1865–1900* (Cambridge: Cambridge University Press, 2000), p. 231.

21. Although scientific consensus in the late 1890s may have aided the ratification of conventions, it did nothing to ensure their enforcement. Howard-Jones, *The Scientific Background of the International Sanitary Conferences*, pp. 28, 55, 99. Harrison, 'Public Health and Medical Research in India', p. 182. Some IMS officials continued to cling to environmental views of disease causation, such as Max von Pettenkofer's groundwater theory that subsoil conditions facilitated cholera spread, while incorporating bacteriological discoveries into their etiological framework. Jeremy D. Isaacs, 'D. D. Cunningham and the Aetiology of Cholera in British India, 1869–1897', *Medical History*, 42 (1998): 301.

22. GOI, Education (Sanitary – Deposit), Mar. 1920, National Archives of India (hereafter NAI). Baldwin 150, 190. See Baldwin for discussion of neoquarantism. Britain officially stopped using quarantine against ships in 1896.

23. *Conférence Sanitaire Internationale de Paris. Procès-Verbaux. 7 Février-3 Avril, 1894* (Paris: Imprimerie Nationale, 1894), p. 377. Monod quoted in Howard-Jones, *International Public Health between the Two World Wars*, 74.

24. 'Foreign Opinion on Cholera', *IMG*, 27 (1892): 288.

25. *Procès-Verbaux*, 1894, pp. 377–81.

26. 'Medical and Sanitary Matters in India', *IMG*, 30 (1895): 349.

27. George C. Kohn (ed.), *Encyclopedia of Plague and Pestilence* (New York: Facts of File, 1995), p. 140. J. N. Hays, *Epidemics and Pandemics: Their Impacts on Human History* (Santa Barbara: ABC-CLIO, 2005), pp. 309, 312. Cf. Richard J. Evans, 'Epidemics and Revolutions: Cholera in Nineteenth-Century Europe', *Past and Present*, 120.1 (1988): 123–46.

28. *Procès-Verbaux*, 1894, pp. 94, 190.

29. Fidler, *International Law and Infectious Diseases*, p. 32.

30. Valeska Huber, 'The Unification of the Globe by Disease? The International Sanitary Conferences on Cholera, 1851–1894', *Historical Journal*, 49.2 (2006): 475. Howard-Jones, *The Scientific Background of the International Sanitary Conferences*, p. 63.

31. *Conference Held at Venice in January 1892 Respecting the Sanitary Regulations of Egypt. Presented to both Houses of Parliament by Command of Her Majesty: June 1892* (London: Harrison and Sons, 1892), pp. 89–90.

32. *Conference*, pp. 89–90. *Procès-Verbaux*, 1894, p. 164.

33. According to international custom, ships were classified as 'infected', 'suspected', or 'healthy' depending on the health status of passengers and crews. For cholera, infected meant that at least one case had occurred within seven days of a ship arriving at its destination or at an intermediate port.

34. 'The New Antiplague Campaign', *IMG*, 42 (1907): 381–2. Rajnarayan Chandavarkar, 'Plague Panic and Epidemic Politics in India, 1896–1914', *Epidemics and Ideas: Essays on the Historical Perception of Pestilence*, Terence Ranger and Paul Slack (eds) (Cambridge: Cambridge University Press, 1992), pp. 206–7.

35. Proceedings of the Government of Bombay (hereafter GOBo), General (General), Feb.–Mar. 1897, IOR/p/5319.

36. GOI, Home (Sanitary), Feb. 1897, IOR/p/5188. The export of hides and skins amounted to over 6 percent of India's export earnings. Harrison, 'Public Health and Medical Research in India', p. 222.

37. GOBo, General (General), Feb.–Mar. 1897, IOR/p/5319. Certain goods, such as grain, hides, and cotton, were thought to harbor the plague bacillus, and thus were subject to greater scrutiny.

38. G. Balachandran, 'Reappraisal: Finance and Politics in Late Colonial India 1917–1947', *South Asia*, 19.1 (1996): 78–9.

39. Azmi Ozcan, *Pan-Islamism: Indian Muslims, the Ottomans and Britain (1877–1924)* (Leiden: Brill, 1997), p. 107.

40. Mark Harrison, *Public Health in British India: Anglo-Indian Preventive Medicine, 1859–1914* (Cambridge: Cambridge University Press, 1994), pp. 136–7. This does not imply a 'communal consciousness' among Muslims on issues related to pilgrimage. In the late nineteenth and early twentieth centuries, there was much variation and oppositional standpoints within Muslim discourse, based on class, regional and ideological differences. Ayesha Jalal, 'Exploding Communalism: The Politics of Muslim Identity in South Asia', *Nationalism, Democracy and Development: State and Politics in India*, Sugata Bose and Ayesha Jalal (eds) (Delhi: Oxford University Press, 1997), pp. 80–91.

41. J. K. Condon, *The Bombay Plague: Being a History of the Progress of Plague in the Bombay Presidency from September 1896 to June 1899* (Bombay: Education Society, 1900), pp. 136–7.

42. Chandavarkar, p. 210.

43. *Conférence Sanitaire Internationale de Venise. 16 Février–19 Mars 1897. Procès-Verbaux* (Rome: Forzani, 1897), pp. 79–80, 117.
44. Harrison, *Public Health in British India*, pp. 136–7.
45. For a discussion of international confidence and health policy, see Donald F. Thompson and Renata P. Louie, 'Cooperative Crisis Management and Avian Influenza: A Risk Assessment Guide for International Contagious Disease Prevention and Risk Mitigation', March 2006: 28–9, http://www.ndu.edu/ctnsp/Def_Tech (accessed 3 April 2009).
46. GOI, Home (Sanitary), Feb. 1897, IOR/p/5288. David Arnold, *Colonizing the Body: State Medicine and Epidemic Disease in Nineteenth-Century India* (Berkeley: University of California Press, 1993), p. 205.
47. *Procès-Verbaux*, 1897, p. 117.
48. GOI, Home (Sanitary), Dec. 1897, IOR/p/5191.
49. Condon, pp. 136–7.
50. *Procès-Verbaux*, 1897, p. 36. GOI, Home (Sanitary), Mar. 1897, IOR/p/5188.
51. *Procès-Verbaux*, 1897, p. 67. No delegates from India were present at the 1897 convention.
52. Ironically, Tilak had earlier volunteered in sanitary search parties in Pune. He was later tried for sedition in relation to the assassination of Walter Rand and was sentenced to imprisonment for 18 months. Myron Echenberg, *Plague Ports: The Global Urban Impact of Bubonic Plague, 1894–1901* (New York: New York University Press, 2007) 63, 66, 68.
53. Arnold, *Colonizing the Body*, 212–15. David Arnold, 'Touching the Body: Perspectives on the Indian Plague, 1896–1900', *Selected Subaltern Studies*, Ranajit Guha and Gayatri Chakravorty Spivak (eds) (Oxford: Oxford University Press, 1988), pp. 405–11. Sapna Patel, 'Responses to an Epidemic: The Attitudes of the Colonial State and the Indian Press to the Plague in Bombay: September 1896 to December 1897', diss., Wellcome Institute, 2005, pp. 23–5.
54. 'Reports on Native Papers: Bombay Presidency', 5.35 (1897), IOR/L/R/5/152.
55. 'Reports on Native Papers' 4.23, 17.9.
56. The Secretary to the Government of India, the Governor of Bombay, and the Secretary of State for India, did support segregating plague cases in separate hospitals with accommodations for sex, caste, and religion, but it was left to local officials to decide whether to make these arrangements. 'Papers Relating to the Outbreak of Bubonic Plague in India; with Statement showing the Quarantine and other restrictions recently placed upon Indian trade, up to March 1897. Presented to Parliament' (London: Eyre and Spottiswoode, 1897), p. 5. 'Lord Sandhurst's Measures Against Plague', *IMG*, p. 32 (1897): 102. *Madras Plague Regulations*, pp. 77–8.
57. GOI, Home (Sanitary) Plague Deposit Collections, vol. 2, 1897–99, NAI.
58. 'Reports on Native Papers', 4.23.
59. 'The Paternal *versus* the Common Sense Plague Policy', *IMG*, 39 (1904): 447–50. Arnold, *Colonizing the Body*, pp. 226, 230–3.
60. Procès-Verbaux 1897, p. 89.
61. GOI, Home (Sanitary), Aug. 1898, IOR/p/5423. Government motives were rumored to include: poisoning people to decrease excess population; spreading disease to deter foreigners from invading India; seizing money and

property; propitiating the plague demon; permanently consolidating British rule in India; and interfering with religious and caste observances to force Christianity on Indians. GOI, Home (Sanitary), Jul. 1900, IOR/p/5883.

62. GOI, Home (Sanitary) Plague Deposit Collections, vol. 2, 1897–99, NAI.

63. GOI, Home (Sanitary), Nov. 1898, IOR/p/5424. Waldemar Haffkine's anti-plague inoculation was made with killed bacilli and his anti-cholera inoculation with attenuated bacilli. 'Inoculation' was used interchangeably with 'vaccine'.

64. Condon, p. 143.

65. Letter from Juan Armand Rüffer, British consul in Alexandria, to Lord Cromer in GOI, Home (Sanitary), Sept. 1900, IOR/p/5884.

66. Proceedings of the Government of Madras (hereafter GOM), Local and Municipal (Plague), Aug. 1900, IOR/p/6027.

67. GOBo, General (General), Sept. 1931, IOR/p/11911. 'Annual Report of the Sanitary Commissioner with the GOI for 1897', p. 197. (In 1920, the Sanitary Commissioner post was renamed to Public Health Commissioner. Annual reports of the Sanitary Commissioner and of the Public Health Commissioner will both be referred to hereafter as 'Annual Report'.)

68. The Government of India recommended, that pilgrims postpone their journey until next season because of Turkey's stringent quarantine rules and the harassment they were likely to experience on arrival. GOM, Local and Municipal (Plague), Nov. 1900, IOR/p/6027. GOM, Local and Municipal (Plague), Jun 1899, IOR/p/5768.

69. Proceedings of the GOI, Sanitary (Plague), April 1899 quoted in *Report of the Committee of the Bengal Chamber of Commerce: From 1st February 1899 to 31st January 1900*, vol. 2 (Calcutta: W. Newman & Company, 1900).

70. Bynum, *Science and the Practice of Medicine in the Nineteenth Century*, p. 145.

71. GOI, Home (Sanitary), Jan. 1894, IOR/p/4555.

72. *Procès-Verbaux*, 1894, pp. 377–81.

73. Into the 1920s, delegates for India maneuvered so as to avoid placing India in a position, in which it might have to violate sanitary conventions and incur undesirable consequences. *Conférence Sanitaire Internrnationale de Paris (10 Mai–21 Juin 1926). Procès-Verbaux* (Paris: Imprimerie Nationale, 1927), pp. 287–9.

74. Acceptance of a convention did not signify formal ratification; signing a convention without ratification allowed both Britain and India more room to maneuver. Whitehall was used to signify British governmental administration.

75. GOI, Home (Sanitary), Jan. 1894, IOR/p/4555. The return of the Liberal Party in Britain, which saw the revival of a Gladstonian approach to empire focusing on free trade and non-interference, may explain why India was allowed to decline ratification. In the late nineteenth century, Britain may also have allowed its colonies more freedom in this area as Britain moved toward closer cooperation with its colonies regarding foreign policy and the sharing of defense costs and resources. Martin Pugh, *The Making of Modern British Politics, 1867–1945* (Oxford: Blackwell Publishers, 2002 [1982]), pp. 101–2. Muriel E. Chamberlain, *The Formation of the European Empires, 1488–1920* (Harlow: Pearson Education, 2000), p. 108.

76. GOI, Home (Sanitary), Jan. 1894, IOR/p/4555.

77. GOI, Home (Sanitary), Oct. 1895, IOR/p/4753.

78. *Procès-Verbaux*, 1894, p. 288.

79. *Procès-Verbaux*, 1894, pp. 278–9.
80. GOI, Home (Sanitary), Oct. 1895, IOR/p/4753.
81. Harrison, *Public Health in British India*, pp. 131–2.
82. The Pilgrim Ships Act of 1895 conformed to the regulations of the Paris convention excluding those provisions antithetical to British and Indian interests. *Report of the Bombay Chamber of Commerce for the Year 1897* (Bombay: Bombay Gazette Steam Printing Works, 1898), pp. 308–11.
83. Proceedings of the Government of Bengal (hereafter GOB), Municipal (Medical), May 1898, West Bengal State Archives.
84. Ibid.
85. GOBo, General (General), Jun. 1909, IOR/p/8302. In 1908, the Government of India finally adhered to the 1903 convention but with reservations.
86. GOI, Home (Sanitary), Dec. 1894 IOR/p/4555.
87. GOI, Home (Sanitary), Oct. 1896, IOR/p/4966. GOI, Home (Sanitary), Nov. 1896, IOR/p/4966.
88. *Proceedings Relating to Quarantine. Vol. VII. 1892–1896*. GOI, Home (Sanitary) November 1896, NAI.
89. GOM, Local and Municipal (Plague), Aug. 1900, IOR/p/6027.
90. Letter from Juan Armand Rüffer, British consul in Alexandria, to Lord Cromer.
91. 'Resolution of the Government of India in the Home Department, Sanitary, No. 227–240, Dated the 3rd February 1898', *The Madras Plague Regulations and Rules* (Madras: Government Press, 1898), pp. 71–2.
92. GOI, Home (Sanitary), Feb. 1900, IOR/p/5882. GOI, Home (Sanitary), April 1900, IOR/p/5883.
93. GOBo, General (General), Sept. 1931, IOR/p/11911. 'Annual Report' (1897): 197.
94. 'Report of the Bombay Plague Committee (for period 1 July 1897–30 April 1898)' (Bombay: Times of India Steam Press, 1898) 8, 33, IOR/v/26/856/3.
95. GOBo, General (General) Nov. 1898, IOR/p/5540.
96. Differing definitions of epidemicity created confusion and conflict between the central and provincial governments and had significant implications for when and how the provisions of international conventions should be applied. The definition of 'epidemic' became problematic in that the application of both international and domestic regulations was often predicated on epidemicity. GOB, Municipal (Medical), Aug. 1897, IOR/p/5172. GOB, Municipal (Medical), Nov. 1897, IOR/p/5173.
97. GOM, Local and Municipal (Plague), Mar. 1900, IOR/p/6027.
98. GOI, Home (Sanitary), Dec. 1897, IOR/p/5191.
99. GOB, Municipal (Medical), Jul. 1899, IOR/p/5630.
100. GOI, Home (Sanitary), Jul. 1898, IOR/p/5423.
101. GOI, Home (Sanitary), Oct., Dec. 1897, IOR/p/5191.
102. GOI, Home (Sanitary), Aug. 1897, IOR/p/5190.
103. GOI, Home (Sanitary), Dec. 1897, IOR/p/5191. Cf. Richard Tuck, *The Rights of War and Peace: Political Thought and the International Order from Grotius to Kant* (Oxford: Oxford University Press, 1999).
104. Alexandra Minna Sternn and Howard Markel, 'International Efforts to Control Infectious Diseases, 1851 to the Present', *JAMA*, 292 (2004): 1474, 1478.
105. 'Annual Report' (1908): 98.

## 2 Plague and Cholera – The Epidemic versus the Endemic

1. Maureen Sibbons, 'Cholera and Famine in British India, 1870–1930', in *Papers in International Development*, No. 14 (Swansea: Centre for Development Studies, 1995): 1.
2. Ira Klein, 'Urban Development and Death: Bombay City, 1870–1914', *Modern Asian Studies*, 20.4 (1986): 744–6.
3. Patrick Zylberman, 'Civilizing the State: Borders, Weak States and International Health in Modern Europe', *Medicine at the Border: Disease, Globalization and Security, 1850 to the Present*, Alison Bashford (ed.) (New York: Palgrave Macmillan, 2006), p. 24.
4. Cf. Ronald Hyam, *Britain's Imperial Century, 1815–1914: A Study of Empire and Expansion* (Lanham: Barnes and Nobles Books, 1993), pp. 285–90. For discussion of gentlemanly capitalism, see P. J. Cain and A. G. Hopkins, *British Imperialism, 1688–2000* (Harlow: Longman, 2002).
5. 'Plague Prevention in Bengal', *Indian Lancet* (16 Sept. 1897): 287–8.
6. Cain and Hopkins, p. 278.
7. Ira Klein, 'Death in India, 1871–1921', *Journal of Asian Studies*, 32 (1973): 645.
8. GOM, Local and Municipal (Plague), Aug. 1900, IOR/p/6027.
9. The Great Depression would later stop or reverse these labor flows. Sugata Bose, *A Hundred Horizons: the Indian Ocean in the Age of Global Empire* (Cambridge: Harvard University Press, 2006), pp. 99–100, 112–13.
10. GOI, Home (Sanitary), Aug. 1897, IOR/p/5190. GOI, Home (Sanitary), Oct. 1897, IOR/p/5191.
11. GOBo, General (General), Mar. 1897, IOR/p/5319.
12. GOI, Home (Sanitary), Apr. 1898, IOR/p/5422.
13. GOI, Home (Sanitary), Aug. 1898, IOR/p/5423.
14. Valeska Huber, 'The Unification of the Globe by Disease? The International Sanitary Conferences on Cholera, 1851–1894', *Historical Journal*, 49.2 (2006): 471. Between 1908 and 1928 (excepting the period of political upheaval in the Hedjaz during World War I and 1925), on average 19,000 pilgrims sailed yearly from Indian ports of which about 13,000 were British Indians. Bose, *A Hundred Horizons*, p. 204.
15. W. J. Simpson, 'Maritime Quarantine and Sanitation in Relation to Cholera', *The Practitioner: A Journal of Therapeutics and Public Health*, 48 (1892): 148–60. In 1893, there were an estimated 13,400 cholera deaths in Mecca. J. N. Hays, *Epidemics and Pandemics: Their Impacts on Human History* (Santa Barbara: ABC-CLIO, 2005), p. 305.
16. E. Wilkinson, 'Report on Inquiries into the Measures for the Sanitary Control of the Hejaz Pilgrimage, 1919', p. 3.
17. Ayesha Jalal, *Self and Sovereignty: Individual and Community in South Asian Islam since 1850* (London: Routledge, 2000), pp. 185–6. Mark Harrison, *Public Health in British India: Anglo-Indian Preventive Medicine, 1859–1914* (Cambridge: Cambridge University Press, 1994), p. 231.
18. GOI, Home (Sanitary), Feb. 1897, IOR/p/5188. Calcutta's port would remain closed to pilgrims traveling to the Hedjaz for another 30 years.
19. GOI, Home (Sanitary), Feb. 1897, IOR/p/5188.
20. GOB, Municipal (Medical), Dec. 1897, IOR/p/5173.

21. 'Papers Relating to the Outbreak of Bubonic Plague in India; with Statement showing the Quarantine and other restrictions recently placed upon Indian trade, up to March 1897. Presented to Parliament' (London: Eyre and Spottiswoode, 1897), pp. 97–8.
22. GOI, Home (Sanitary), Feb. 1897, IOR/p/5188.
23. GOI, Home (Sanitary), Nov. 1897, IOR/p/5191.
24. GOBo, General (General), Feb. 1898, IOR/p/5540.
25. GOI, Home (Sanitary), Feb. 1897, IOR/p/5188. GOI, Home (Sanitary), Apr. 1898, IOR/p/5422.
26. GOI, Home (Sanitary), Jun. 1898, IOR/p/5423.
27. GOI, Home (Sanitary), Apr. 1901, IOR/p/6118.
28. GOI, Home (Sanitary), Jul. 1903, IOR/p/6583.
29. GOI, Home (Sanitary), Mar. 1905, IOR/p/7058. GOI, Education (Sanitary), Aug. 1913, IOR/p/9200.
30. W. F. Bynum, 'Policing Hearts of Darkness: Aspects of the International Sanitary Conferences', *History and Philosophy of the Life Sciences*, 15 (1993): 431. Alexandra Minna Stern and Howard Markel, 'International Efforts to Control Infectious Diseases, 1851 to the Present', *JAMA*, 292 (2004): 1476.
31. GOI, Home (Sanitary), Jul 1904, IOR/p/6816.
32. GOB, Municipal (Medical), Nov. 1908, IOR/p/7564.
33. F. Norman White, *Twenty Years of Plague in India with Special Reference to the Outbreak of 1917–18* (Calcutta: Superintendent Government Printing, 1918), p. 14.
34. GOI, Home (Sanitary), Sept. 1902, IOR/p/6351.
35. GOI, Home (Sanitary), Oct. 1906, IOR/p/7326. Deratisation was the contemporary term for eliminating rats from a defined space. GOI, Home (Sanitary) Plague Deposit Collections, vol. 3, 1900–02, NAI.
36. British India was the only country participating in the OIHP to affirm the efficacy of the anti-plague vaccine in reducing plague mortality. G. Abt, *Vingt-Cinq Ans d'Activité de L'Office International D'Hygiène Publique, 1909–1933* (Paris: Office International D'Hygiène Publique, 1933), p. 77.
37. GOI, Home (Sanitary) Feb. 1900, IOR/p/5882.
38. Shortly after the cholera epidemic of 1892, Waldemar Haffkine began trials of his anti-cholera inoculation in India, administering 42,445 inoculations from 1893 to 1895. Trials demonstrated a reduction in cholera morbidity and mortality among the inoculated; however, Haffkine doubted the efficacy of the inoculation in addressing endemic cholera. W. M. Haffkine, 'A Lecture on Vaccination Against Cholera', *Indian Lancet* (16 Feb. 1896): 170–3. A. Crombie, 'Haffkine's Anti-Choleraic Inoculation', *Medical Reporter* (16 Sept. 1895): 173. Ilana Löwy, 'Producing a Trustworthy Knowledge: Early Field Trials of Anticholera Vaccines in India', *Vaccinia, Vaccination, Vaccinology: Jenner, Pasteur and their Successors*, Stanley A. Plotkin and Bernardino Fantini (eds) (Paris: Elsevier, 1996), p. 121.
39. GOI, Education (Sanitary), Feb. 1916, IOR/p/9946.
40. GOI, Education (Sanitary), Feb. 1920, IOR/p/10830.
41. WHO, ARC001, Records of the Office International d'Hygiène Publique, 1907–1946, Microfilm: T12 7-8. GOI, Home (Sanitary), Jun. 1908, IOR/p/7886. 'The Office International d'Hygiene Publique', *IMG*, 65 (1930): 582. GOI, Education (Health), Jun. 1930, IOR/p/11862.

42. A. R. Wellington, *Hygiene and Public Health in India. Report on Conditions Met with During the Tour of the League of Nations Interchange of Health Officers* (Kuala Lumpur: Federated Malay States Government Press, 1929), p. 28.
43. Norman Howard-Jones, *The Scientific Background of the International Sanitary Conferences, 1851–1938* (Geneva: World Health Organization, 1975), p. 97.
44. GOI, Education, Health and Lands (Health), Jan. 1931, B. 193-7, NAI.
45. Bose, *A Hundred Horizons*, p. 208.
46. GOI, Education, Health and Lands (Health), Jan. 1931, B. 193-7, NAI.
47. Saurabh Mishra, *Pilgrimage, Politics, and Pestilence: The Haj from the Indian Subcontinent, 1860–1920* (New Delhi: Oxford University Press, 2011), p. 15.
48. David Arnold, 'Cholera and Colonialism in British India', *Past and Present*, 113 (1986): 138–9.
49. In the 1930s, 1–1.5 million inoculations were given annually in Bengal to prevent the disease spreading beyond the province's borders. 'Cholera and Bengal', *IMG*, 68 (1933): 521–2.
50. 'Report of the Pilgrim Committee, Madras: 1915' (Simla: Government Monotype Press, 1916) 11, IOR/v/26/844/4.
51. League of Nations (hereafter LON), 'Minutes of the Tenth Session of the Health Committee, April 26–27, 1927', pp. 8–9.
52. *Proceedings of the Third All-India Sanitary Conference held at Lucknow in January 1914, and Resolution of the Government of India on Sanitation in India* (London: His Majesty's Stationary Office, 1914), p. 14, Appendix 14.
53. Leonard Rogers, 'Progress in Control of Cholera in India by Inoculation of Pilgrims', *British Medical Journal*, 2.4796 (1952): 1219. Arnold, 'Cholera and Colonialism in British India', p. 147.
54. 'Report (1939) of the Sub-Committee appointed by the Central Advisory Board of Health to examine the possibility of introducing a system of compulsory inoculation of pilgrims against cholera' (Simla: Government of India Press, 1940) 2-4, IOR/v/26/844/7. Arnold, 'Cholera and Colonialism in British India', p. 149. In 1932, 97 percent of pilgrims coming from southern India were inoculated against cholera. Abt 83–4.
55. 'Report (1939)'. Rogers 1219.
56. The Central Advisory Board was established in 1937 to coordinate provincial activities, establish greater uniformity in public health measures, and ensure coordination between central and provincial governments in health matters of mutual concern. 'The Cholera Danger in India: Melas and the Spread of Cholera', *IMG*, 74 (1939): 489. GOI, Education, Health and Lands (Health) 29-11/40-H. 1940, NAI. Provincial elections had brought the Congress government to office in eight provinces by 1938. This may have influenced the decision to maintain provincial discretion in mandating the inoculation of pilgrims.
57. 'Preventive Medicine a Factor in Empire Building', *IMG*, 39 (1904): 382. Michael Worboys, 'Manson, Ross, and Colonial Medical Policy: tropical medicine in London and Liverpool, 1899–1914', *Disease, Medicine, and Empire: Perspectives on Western Medicine and the Experience of European Expansion*, Roy MacLeod and Milton Lewis (eds) (London: Routledge, 1988), pp. 26–7.
58. Cf. Bernard Cohn, *Colonialism and its Forms of Knowledge: The British in India* (Princeton: Princeton University Press, 1996).

180   *Notes*

. Harrison, *Public Health in British India*, 165. Cf. Richard Drayton, *Nature's Government: Science, Imperial Britain, and 'Improvement' of the World* (New Haven: Yale University Press, 2000) 220. Cf. Douglas Haynes, *Imperial Medicine: Patrick Manson and the Conquest of Tropical Disease* (Philadelphia: The University of Pennsylvania Press, 2001).
60. 'Medical Research in India', *IMG*, 52 (1917): 288.
61. 'Medical and Sanitary Matters in India', *IMG*, 30 (1895): 349–50.
62. 'Medical Research in India', p. 165. David Arnold, *Science, Technology and Medicine in Colonial India* (Cambridge: Cambridge University Press, 2000), p. 144. Harrison, *Public Health in British India*, p. 165. Deepak Kumar, 'Perceptions of Public Health: A Study in British India', *Maladies, Preventives and Curatives: Debates in Public Health in India*, Amiya Kumar Bagchi and Krishna Soman (eds) (Kolkata: Tulika Books, 2005), p. 47.
63. Michael Worboys, 'British Colonial Medicine and Tropical Imperialism: A Comparative Perspective', *Dutch Medicine in the Malay Archipelago, 1816–1942*, A. M. Luyendijk-Elshout (ed.) (Amsterdam: Rodopi B.V., 1989), p. 159.
64. Apart from the Central Research Institute at Kasauli, all laboratories were under the direction of provincial governments or Pasteur associations. O. P. Jaggi, *Medicine in India: Modern Period* (New Delhi: Oxford University Press, 2000), p. 256. By 1937, this laboratory had produced 41 million doses of Haffkine's plague vaccine for use within India and abroad. E. W. C. Bradfield, 'Notes on Medical and Public Health Organisation in the Bombay Presidency', *Intergovernmental Conference of Far-Eastern Countries on Rural Hygiene: Preparatory Papers relating to British India* (Geneva: League of Nations, 1937), pp. 213–15.
65. 'Research in Tropical Diseases in India', *IMG*, 40 (1905): 307.
66. GOM, Local and Municipal (Plague), Dec. 1898, IOR/p/3788.
67. GOB, Municipal (Medical) Dec. 1896, IOR/p/4850.
68. Arnold, *Science, Technology and Medicine in Colonial India*, p. 144.
69. GOI, Home (Sanitary), Mar. 1902, IOR/p/6350. 'The Plague Commission Report', *IMG*, 35 (1900): 141–2.
70. GOBo, General (Medical), Jun. 1905, IOR/p/7189.
71. 'Travaux de la Commission Anglaise de la Peste aux Indes', *Bulletin de l'Office International d'Hygiène Publique* (Jan. 1909): 94, 105.
72. *The Proceedings of the Second All-India Sanitary Conference held at Madras in November 1912*, vol. 1 (Simla: Government Central Branch Press, 1913), p. 9. A. J. H. Russell, 'A Note on the Central Government's Health Organization and Associated Institutions and Organisations Concerned with Public Health', *Intergovernmental Conference of Far-Eastern Countries on Rural Hygiene: Preparatory Papers relating to British India* (Geneva: League of Nations, 1937) 21. 'Indian Medical Research', *IMG*, 73 (1938): 230.
73. GOI, Education (Sanitary), May 1913, IOR/p/9199.
74. *The Proceedings of the First All-India Sanitary Conference held at Bombay on 13th and 14th November 1911* (Calcutta: Superintendent Government Printing, 1912), pp. 1–2.
75. 'Report of the Committee on the Organisation of Medical Research under the Government of India' (Calcutta: Government of India Central Publication Branch, 1929) 10–15, IOR/v/26/850/7. This shortage of qualified research workers led to the creation, with the support of the Rockefeller Foundation

and industry associations, of the Calcutta School of Tropical Medicine and Hygiene in 1922, established to train Indians to investigate major communicable diseases in India. *Annals of the Indian Tea Association for 1915* (Calcutta: Indian Tea Association 1916), pp. 273, 281–2.

76. 'Report of the Committee on the Organisation of Medical Research', pp. 10–15.
77. Arnold, *Science, Technology and Medicine in Colonial India*, p. 186.
78. The League's Health Organisation acted as an information clearinghouse for national vital statistics and epidemiological information, helped organize efforts to combat epidemic diseases, and coordinated scientific research work. J. D. Graham, 'International Aspects of Disease with Special Reference to Quarantine', *Far Eastern Association of Tropical Medicine. Transactions of the Seventh Congress, British India 1927*, vol. 1 (Calcutta: Thacker's Press, 1928), pp. 466–8. League of Nations Health Organization, *International Health Year-Book, 1925*, vol. 2 (Geneva: League of Nations, 1926), p. 610. Sheldon Watts, 'British Development Policies and Malaria in India 1897–c. 1929', *Past and Present*, 165 (1999): 177–8.
79. LON, 12B Health, R991, 62193/57535 and 58870.
80. LON, CH 687, 'Resolutions Adopted by the Advisory Council of the Eastern Bureau at its Session Held in New Delhi From the 26th to the 29th Dec, 1927', p. 3.
81. J. D. Graham, 'La Peste dans l'Inde Britannique', *Bulletin de l'Office International d'Hygiène Publique*, 22.11 (1930): 2088–9. GOI, Education, Health and Lands (Health) Aug. 1929, B. 94–95, NAI.
82. 'Annual Report of the All-India Institute of Hygiene and Public Health for the Year 1934', *IMG*, 70 (1935): 535–6.
83. Rajnarayan Chandavarkar, 'Plague Panic and Epidemic Politics in India, 1896–1914', *Epidemics and Ideas: Essays on the Historical Perception of Pestilence*, Terence Ranger and Paul Slack (eds) (Cambridge: Cambridge University Press, 1992), p. 209.
84. GOI, Home (Sanitary), Dec. 1897, IOR/p/5191.
85. *Conférence Sanitaire Internationale de Venise. 16 Février–19 Mars 1897. Procès-Verbaux* (Rome: Forzani, 1897), p. 89.
86. Frank G. Clemow, *The Geography of Disease* (Cambridge: University Press, 1903), p. 349. W. M. Haffkine, 'On the Present Methods of Combating the Plague', *Proceedings of the Royal Society of Medicine* (Epidemiological Section) 1.1 (1908): 79–80.
87. Nield Cook, 'Plague Precautions', *Indian Lancet* (16 May 1898): 479–82. Haffkine also developed an anti-plague inoculation and began testing it in 1897.
88. W. M. Haffkine and Surgeon-Major Lyons, 'Joint Report on the Epidemic of Plague in Lower Damaun, Portuguese India, and on the Effect of Preventive Inoculation There', *Indian Lancet* 8.12 (1897): 594–5.
89. A. B. Fry, 'Le Choléra dans le Bengale, dans le Passé et à l'Heure Actuelle', *Bulletin de l'Office International d'Hygiène Publique*, 18.3 (1926): 299.
90. Clemow, p. 349. The Government of India responded to negative criticism domestically and internationally by increasing funds allocated to provincial governments for sanitation, but grants only increased significantly after 1908, when revenue surpluses permitted increased expenditure. Mark Harrison, 'Public Health and Medicine in British India: an Assessment of the British Contribution', *Bulletin of the Liverpool Medical History Society*, 10 (1998): 44.

91. J. A. Turner, *Sanitation in India* (Bombay: Times of India, 1914), p. 498.

92. Ibid., pp. 554–5.

93. Haffkine, 'On the Present Methods of Combating the Plague', pp. 79–80.

94. 'Presidential Address in Public Health', *IMG*, 30 (1895): 18.

95. Haffkine and Lyons, pp. 594–5. David Arnold, 'Touching the Body: Perspectives on the Indian Plague, 1896–1900', *Selected Subaltern Studies*, Ranajit Guha and Gayatri Chakravorty Spivak (eds) (Oxford: Oxford University Press, 1988), p. 418.

96. *Procès-Verbaux 1897*, p. 89.

97. Clemow, p. 349.

98. Wm. Wesley Clemesha, *Plague, From the Sanitarian's Point of View* (Calcutta: Baptist Mission Press, 1903), pp. 52–6.

99. 'Major Bannerman, I. M. S., on the Results of Four Years' Inoculations Against Plague', *IMG*, 36 (1901): 181.

100. Anil Kumar, *Medicine and the Raj: British Medical Policy in India, 1835–1911* (Walnut Creek: Altamira Press, 1998), p. 203.

101. William Glen Liston, 'The Cause and Prevention of the Spread of Plague in India. A Lecture Delivered before the Bombay Sanitary Association on 11th December 1907'.

102. 'The Plague Commission Report', pp. 141–2.

103. In Bengal evacuation continued to be the preferred method of combating plague. 'Sanitation in the Two Bengals', *IMG*, 46 (1911): 361.

104. Claire Hooker, 'Sanitary Failure and Risk: Pasteurisation, Immunisation and the Logics of Prevention', *Contagion: Epidemics, History and Culture from Smallpox to Anthrax*, Alison Bashford and Claire Hooker (eds) (Annandale: Pluto Press Australia, 2002 [2001]), pp. 129, 144.

105. Worboys, 'Manson, Ross, and Colonial Medical Policy', pp. 26–7, 32–3. Haynes, pp. 9–10, 154.

106. Cook, pp. 479–82.

107. Haffkine, 'On the Present Methods of Combating the Plague', pp. 79–80.

108. *Intergovernmental Conference of Far-Eastern Countries on Rural Hygiene*, pp. 288–9.

109. I. J. Catanach, '"Fatalism"? Indian Responses to Plague and Other Crises', *Asian Profile*, 12.2 (1984): 191. Löwy, p. 123.

110. GOI, Education (Sanitary), Mar. 1918,IOR/p/10364.

111. J. Taylor, 'Rural Plague in India', *Intergovernmental Conference of Far-Eastern Countries on Rural Hygiene: Preparatory Papers relating to British India* (Geneva: League of Nations, 1937), pp. 86–8. C. M. Ganapathy, 'Public Health Administration in Madras Presidency', *Intergovernmental Conference of Far-Eastern Countries on Rural Hygiene. Preparatory Papers Relating to British India* (Geneva: League of Nations, 1937), pp. 288–9.

112. *Proceedings of the Second All-India Sanitary Conference*, vol. 1, p. 109.

113. *Proceedings of the Third All-India Sanitary Conference*, p. 236.

114. W. Glen Liston, 'Plague Preventive Measures', *The Proceedings of the Second All-India Sanitary Conference held at Madras in November 1912*, vol. 3 (Simla: Government Central Branch Press, 1913), p. 100.

115. *Proceedings of the Third All-India Sanitary Conference*, Appendix 8–9. Liston, 'Plague Preventive Measures', p. 104. 'Notes Epidemiologiques sur la Peste dans l'Inde', *Bulletin de l'Office International d'Hygiène Publique* (May 1924): 585.

116. A. J. H. Russell, 'Périodicité du Cholera dans l'Inde', *Bulletin de l'Office International d'Hygiène Publique*, 17.8 (1925): 901.
117. R. C. Agarwal, *Constitutional Development and National Movement of India* (New Delhi: S. Chand, 1991), pp. 203–4.
118. GOI, Education (Sanitary), Aug. 1920, IOR/p/10830.
119. F. Norman White, *The Prevalence of Epidemic Disease and Port Health Organisation and Procedure in the Far East* (Geneva: League of Nations, 1923), p. 143.
120. White, *Twenty Years of Plague in India*, p. 16. Rat immunity was more prevalent in areas hardest hit by plague.
121. 'Annual Report' (1925): 175.
122. 'Annual Report' (1926): 212.
123. A. J. H. Russell, 'Cholera in India', Far Eastern Association of Tropical Medicine. *Transactions of the Ninth Congress held at Nanking, 1934*, vol. 1 (Nanking: National Health Administration, 1935?), pp. 390–4.
124. 'Cholera and Bengal', p. 521.
125. *Proceedings of the Third All-India Sanitary Conference*, Appendix 12, 15–16. In the 1920s, provincial public health departments increased their public health spending to between 20 and 70 percent of annual total income. Most of it was directed at waste disposal and improving water supplies and drainage. 'Annual Report' (1925–1932).
126. 'Cholera and Bengal', pp. 521–2.
127. A. L. Hoops, *Present Day Public Health in India: A Report on the League of Nations Interchange of Health Officers in India (1st January–18th February, 1928)* (London: John Bale, Sons, and Danielsson), p. 53.
128. 'Cholera and Bengal', pp. 521–2.
129. *Ninth Conference of Medical Research Workers held at Calcutta from 30th November to 5th December 1931* (Simla: Government of India Press, 1932), p. 32.
130. 'Annual Report' (1926): 209–10.
131. 'Annual Report of the Director of Public Health, Madras' (1931): 10.
132. Jane Samson, *Race and Empire* (Harlow: Pearson Education Limited, 2005), p. 85. David Ludden, 'Orientalist Empiricism: Transformations of Colonial Knowledge', *Orientalism and the Postcolonial Predicament: Perspectives on South Asia*, Carol A. Breckenridge and Peter van der Veer (eds) (Philadelphia: University of Pennsylvania Press, 1993), pp. 251–2.
133. GOM, Local (Public Health), Jun., Dec. 1936, IOR/p/12089. GOM, Local (Public Health), Sept. 1930, IOR/p/11853.
134. GOM, Local (Public Health), Jan. 1933, IOR/p/12006.
135. GOM, Local (Public Health), Dec. 1936, IOR/p/12089. GOM, Local (Public Health), Jun. 1936, IOR/p/12089. In the Madras Presidency, two to three tons of printed material, in five languages, was issued annually. The government utilized various media and modes of disseminating health information, including lectures, pamphlets, lantern demonstrations, theater, posters, lessons at village schools, informal village chats, and local newspapers. Hoops 56.
136. 'League of Nations Health Delegation', *Modern Review*, 43.2 (1928): 235.
137. David Arnold, 'Public Health and Public Power: medicine and hegemony in colonial India', *Contesting Colonial Hegemony: State and Society in Africa and India*, Dagmar Engels and Shula Marks (eds) (London: British Academic Press, 1994), p. 137.

138. 'Rats and Plague', *Modern Review*, 43.6 (1928): 765–6.
139. 'Annual Report' (1891–1940).
140. A. J. H. Russell, 'Plague in India', *Far Eastern Association of Tropical Medicine. Transactions of the Ninth Congress held at Nanking*, 1934, vol. 2 (Nanking: National Health Administration, 1935?), pp. 725–7.
141. 'Annual Report' (1929): 301.
142. 'Annual Report' (1900): 123.
143. GOI, Home (Sanitary), Feb. 1900, IOR/p/5882. GOI, Home (Sanitary), Apr. 1906, IOR/p/7325.
144. *Report of the Bombay Chamber of Commerce for the Year 1899* (Bombay: Bombay Gazette Steam Printing Works, 1900), p. 59.
145. Kathleen J. Tierney, 'Toward a Critical Sociology of Risk', *Sociological Forum*, 14.2 (1999): 218, 220, 233.
146. *Statistical Abstract Relating to British India from 1881–82 to 1890–91* (London: Eyre and Spottiswoode, 1892). East India: Accounts and Estimates, 1901–1902 (London: Darling and Son, 1901). *East India: Accounts and Estimates, 1912–1913* (London: Darling and Son, 1912). *East India: Accounts and Estimates, 1921–1922* (London: Eason and Son, 1921). *East India: Accounts and Estimates, 1932–1933* (London: His Majesty's Stationery Office, 1932). *Return of the Budget of the Governor General of India in Council for 1939–40* (London: His Majesty's Stationery Office, 1939).
147. 'Annual Report' (1930): 2.
148. Provincial governments built rat-proof grain storage in villages and towns extensively engaged in grain trading and utilized fumigation, trapping, and poison-baiting in areas where plague tended to become endemic. Taylor, pp. 86, 92.
149. Even at the provincial level, health officials continued to disagree into the late 1930s over whether to adopt a treatment- and technologically-centered approach or a general sanitation approach. William C. Summers, 'Cholera and Plague in India: The Bacteriophage Inquiry of 1927–1936', *Journal of the History of Medicine and Allied Sciences*, 48.3 (1993): 300.
150. Charles E. Rosenberg, 'Framing Disease', *Explaining Epidemics and Other Studies in the History of Medicine* (Cambridge: Cambridge University Press, 1992).
151. Ronald Inden, *Imagining India* (London: Hurst & Company, 2000). Cf. Edward W. Said, *Orientalism* (New York: Vintage Books, 1978).
152. Arnold, 'Public Health and Public Power', pp. 146–7.
153. Despite a decreased risk of pandemic cholera and plague in India, health authorities around the world still considered India to be the main reservoir of cholera and plague well into the 1930s. 'Annual Report of the Public Health Commissioner with the Government of India for 1937, vol. 1', *IMG*, 74 (1939): 774.

## 3  Malaria – India's True Plague

1. Ronald Ross, 'Indian Fevers', *Philosophies* (London: John Murray, 1911), p. 21. Ross wrote this poem during his appointment in Bangalore from 1890–3.
2. *Health Organisation in British India* (Calcutta: Thacker's Press, 1928), p. 37.
3. LON, 8C Health, R6173, 34241, 1561. GOI, Education and Health (Sanitary). Sept. 1922, B. 35, NAI.

4. J. A. Sinton, *What Malaria Costs India* (Delhi: Government of India Press, 1956 [1939]), pp. 4–22. In early life, infants died mostly from the indirect effects of malaria such as premature birth and malnutrition; whereas in later childhood, children died by direct or secondary infections. Adults tended to die from secondary illnesses.

5. Paul F. Russell, 'Malaria in India: Impressions from a Tour', *The American Journal of Tropical Medicine*, 16 (1936): 655.

6. 'View of the Government of India on the Memoranda submitted by Sir Ronald Ross and Colonel W. G. King, I. M. S., regarding the Prevention of Malaria in India', in GOI, Education (Sanitary), Jan. 1912, A. 32–3, NAI.

7. Ira Klein, 'Development and Death: Reinterpreting Malaria, Economics And Ecology in British India', *Indian Economic and Social History Review*, 38.2 (2001): 172–3. Elizabeth Whitcombe, 'The Environmental Costs of Irrigation in British India: Waterlogging, Salinity, Malaria', *Nature, Culture, Imperialism: Essays on the Environmental History of South Asia*, David Arnold and Ramachandra Guha (eds) (Delhi: Oxford University Press, 1995), p. 239.

8. 'The Economic Factor in Tropical Diseases', *IMG*, 57 (1922): 342–3.

9. J. A. Sinton and Raja Ram, *Man-Made Malaria in India* (Simla: Government of India Press, 1938), pp. 1–4.

10. W. A. P Schüffner, 'Le Paludisme aux Indes Britanniques', *Compte-Rendu du Deuxième Congrès International du Paludisme et de la Célébration du Cinquantenaire del Découverte de Laveran* (Alger: Institut Pasteur, 1931), pp. 500–2. Ira Klein, 'Malaria and Mortality in Bengal, 1840–1921', *Indian Economic and Social History Review*, 9.2 (1972): 132–40, 160. Sugata Bose, *Agrarian Bengal: Economy, Social Structure and Politics, 1919–1947* (Cambridge: Cambridge University Press, 1986), pp. 37–8, 45.

11. Sugata Bose, *Peasant Labour and Colonial Capital: Rural Bengal since 1770* (Cambridge: Cambridge University Press, 1993), pp. 24–5.

12. Arabinda Samanta, *Malarial Fever in Colonial Bengal, 1820–1939: Social History of an Epidemic* (Kolkata: Firma KLM, 2002), pp. 197–9.

13. *Twelfth Conference of Medical Research Workers held at Calcutta from 26th November to 1st December 1934* (Simla: Government of India Press, 1935), p. 145.

14. Samanta, p. 195.

15. Russell, 'Malaria in India', p. 654.

16. B. R. Tomlinson, *Political Economy of the Raj, 1914–1947: The Economics of Decolonization in India* (London: Macmillan Press, 1979), p. 2.

17. Sinton, *What Malaria Costs India*, pp. 37–9, 70.

18. LON, 8C Health, R6164, 903/903. LON, 8C Health, R6173, 34241, 1561. GOI, Education and Health (Sanitary). Sept. 1922, B. 35, NAI.

19. A. J. H. Russell, 'A Note on the Central Government's Health Organization and Associated Institutions and Organisations Concerned with Public Health', *Intergovernmental Conference of Far-Eastern Countries on Rural Hygiene: Preparatory Papers relating to British India* (Geneva: League of Nations, 1937), p. 21.

20. S. R. Christophers, *Malaria in the Duars* (Simla: Government Monotype Press, 1911), pp. 21–2.

21. Russell, 'Malaria in India', p. 654.

22. Debates about the prophylactic efficacy of quinine continued into the 1930s until further experiments proved that daily doses of quinine throughout the malaria transmission season could be effective at prevention. *The Treatment of*

*Malaria: Study of Synthetic Drugs, as compared with Quinine, in the Therapeutics and Prophylaxis of Malaria* (Geneva: League of Nations, 1937), p. 124. Today treatment is seen as a critical component of malaria prevention. In twentieth-century India, public health officials made a sharper distinction between the two.

23. V. R. Muraleedharan, '"Cinchona" Policy in British India: The Critical Early Years', *Maladies, Preventives and Curatives: Debates in Public Health in India*, Amiya Kumar Bagchi and Krishna Soman (eds) (Kolkata: Tulika Books, 2005), p. 39.

24. GOBo, General (Medical), Aug., Nov. 1894, IOR/p/4657. Because of malaria's impact on workers, the GOI had also invited private employers, zamindars, and indigo and tea planters to partake of the distribution system for quinine in their provinces.

25. *Paludism, Being the Transactions of the Committee for the Study of Malaria in India* (1911): 10–12. District officers, village schoolmasters, and vaccinators were also authorized to sell quinine packets. Critics argued that the packaging of quinine into single doses misled the public into believing that single doses could be efficacious. E. Wilkinson, 'A Revised Scheme for the Distribution of Quinine by Government', in *Proceedings of the Imperial Malaria Conference held at Simla in October 1909* (Simla: Government Central Branch Press, 1910), p. 79.

26. GOB, Municipal (Medical), Sept. 1895, IOR/p/4740.

27. 97 percent of the world's supply of quinine came from Java with Bengal and Madras producing 2.5 percent. About 200,000 pounds of quinine were being consumed annually in India by the end of the 1930s. C. F. Strickland, *Quinine and Malaria in India* (London: Oxford University Press, 1939), pp. 4–5, 9. 'Summary Report of the Inaugural Meeting of the Central Advisory Board of Health', p. 34.

28. Charles A. Bentley, *Report of an Investigation into the Causes of Malaria in Bombay and the Measures Necessary for its Control* (Bombay: Government Central Press, 1911), pp. 54–5. The GOI defined 'epidemic' as an outbreak where mortality registered above the quinquennial average.

29. GOB, Municipal (Medical), May 1907, IOR/p/7579.

30. The GOI manufactured quinine in government factories primarily from imported cinchona bark.

31. GOB, Municipal (Medical), Nov. 1908, IOR/p/7564.

32. The world's supply of quinine had stabilized at 1,000,000 pounds annually, enough to treat only 30 million people. In India alone, 100 million were afflicted by malaria annually. Wilkinson, 'A Revised Scheme for the Distribution of Quinine by Government', p. 107.

33. GOI, Education (Sanitary), Jun. 1920, IOR/p/10830. GOI, Education (Sanitary), Oct. 1921, IOR/p/11042.

34. GOB, Municipal (Medical), Apr. 1911, IOR/p/8687.

35. LON, 8C Health, R5961, 27535/21313, 'Note on the Quinine Position in India', 1932:1.

36. 'A Visit to the Cinchona Plantations, Bengal', *IMG*, 47 (1912): 289–91.

37. GOBo, General (Medical), Feb. 1911, IOR/p/8830.

38. Jun 1, 1900 letter from D. Prain, Superintendent of the Royal Botanic Garden, Calcutta, to the Under-Secretary for the Government of Bengal. GOB, Municipal (Medical), Mar. 1901, IOR/p/6098.

39. Ibid.
40. *Paludism, Being the Transactions of the Committee for the Study of Malaria in India*, 3 (1911): 13.
41. 'Annual Report' (1908): 98.
42. Edwin R. Nye and Mary E. Gibson, *Ronald Ross: Malariologist and Polymath: A Biography* (London: Macmillan, 1997), pp. 80, 162–4.
43. Ronald Ross, 'A Memorandum on the Present Position of Malaria-Prevention in India', in GOI, Education (Sanitary), Jan. 1912, IOR/p/8946. Ross wrote in his memoirs that he had not 'received any recognition that I am aware of from the Government of India, or even from my old service. I have never been consulted even on my own subject [malaria] by that Government or by the India Office; never been placed on any committee connected with it; never been asked officially for my advice; never received any Indian honour, honorary promotion, or reward'. Ronald Ross, *Memoirs, with a Full Account of the Great Malaria Problem and its Solution* (New York: E.P. Dutton, 1923), p. 357.
44. GOI, Education (Sanitary), Jul. 1919, IOR/p/10588. The GOI occasionally supplied provinces with free quinine to assist provincial quinine distribution. Russell, 'A Note on the Central Government's Health Organization', pp. 23–4.
45. GOI, Home (Sanitary), May 1910, IOR/p/8444.
46. 'The Suppression of Plague and Malaria in India', *IMG*, 46 (1911): 429–30.
47. GOI, Home (Sanitary), May 1910, IOR/p/8444.
48. Mark Harrison, *Public Health in British India: Anglo-Indian Preventive Medicine, 1859–1914* (Cambridge: Cambridge University Press, 1994), pp. 158–63.
49. Michael Worboys, 'Manson, Ross, and Colonial Medical Policy: Tropical Medicine in London and Liverpool, 1899–1914', *Disease, Medicine, and Empire: Perspectives on Western Medicine and the Experience of European Expansion*, Roy MacLeod and Milton Lewis (eds) (London: Routledge, 1988), p. 21.
50. Harrison, *Public Health in British India*, pp. 161–2.
51. J. T. W. Leslie, 'An Address on Malaria in India', *Lancet*, 174 (1909): 1483–4.
52. Ross, 'A Memorandum on the Present Position of Malaria-Prevention in India'.
53. Ibid.
54. Patrick Hehir, *Malaria in India* (London: Oxford University Press, 1927), pp. 425–6. Hehir had studied malaria in India for 25 years.
55. WHO Archives, ARC007, Collection on Parasitology of the Documentation Centre Communicable Diseases, WHO Headquarters: India Malaria, 'Records of the Malaria Survey of India' (1932?): 157–8. G. Covell, *Malaria in Bombay* (Bombay: Government Central Press, 1928), p. 9.
56. W. G. King, 'The Prevention of Malaria in India' in GOI, Education (Sanitary), Jan. 1912, IOR/p/8946.
57. Ibid.
58. C. A. Bentley, 'Propagande en faveur de la quinisation au Bengale' Bulletin de l'Office International d'Hygiène Publique (Oct. 1913): 1864. C. A. Bentley, 'Quinine Propaganda', *Proceedings of the Third Meeting of the General Malaria Committee held at Madras November 18, 19 and 20, 1912* (Simla: Government Central Branch Press, 1913). Cf. Frank M. Snowden, *The Conquest of Malaria: Italy, 1900–1962* (New Haven: Yale University Press, 2006).

59. Dakshina R. Ghosh, 'How to Fight Malaria in Our Villages', *Modern Review*, 19.2 (1916): 199–201.
60. 'The Need for a Public Health for India', *Indian Medical Gazette*, 62 (1927): 576.
61. Malcolm Watson, 'Observations on Malaria Control, with Special Reference to the Assam Tea Gardens, and Some Remarks on Mian Mir, Lahore Cantonment', *Transactions of the Royal Society of Tropical Medicine and Hygiene*, 18.4 (1924), p. 152.
62. Watson, pp. 155, 159.
63. GOI, Education (Sanitary), Jan. 1912, IOR/p/8946.
64. 'View of the Government of India on the Memoranda'.
65. 'Annual Report' (1926): 210. Bonification had been initially popularized in Italy as a means of controlling malaria. Hughes Evans, 'European Malaria Policy in the 1920s and 1930s: The Epidemiology of Minutiae', *Isis*, 80 (1989), p. 43.
66. C. A. Bentley, 'Une nouvelle conception du paludisme', *Bulletin de l'Office International d'Hygiène Publique* (Oct. 1910): 1863.
67. Kabita Ray, 'Press and the Problem of Medical Relief in Colonial Bengal, 1921–1947', *Maladies, Preventives and Curatives: Debates in Public Health in India*, Amiya Kumar Bagchi and Krishna Soman (eds) (New Delhi: Tulika Books, 2005), p. 68.
68. Sailaj Lal Chatterjee, 'Anti-Mosquito Measures', *Calcutta Municipal Gazette*, 3.19 (1926): 815. This was one of several hundred local anti-malaria cooperative societies organized by Indians to fight malaria.
69. 'The Menace of Malaria', *Modern Review*, 40.4 (1926): 463.
70. 'Records of the Malaria Survey of India', p. 154.
71. Ibid., pp. 157–8.
72. Ibid., p. 159.
73. *The Proceedings of the Second All-India Sanitary Conference held at Madras in November 1912*, vol. 1 (Simla: Government Central Branch Press, 1913), pp. 1, 8.
74. *Health Organisation in British India*, p. 16.
75. GOI, Education (Sanitary), Jan. 1912, IOR/p/8946. At the time, the GOI allocated about 3 percent per capita of total revenue to the medical department. 'View of the Government of India on the Memoranda'.
76. J. W. D. Megaw, *Confidential – Further Note on the Formation of a Public Health Board* (Simla: Government of India Press, 1932), p. 14.
77. Russell, 'Malaria in India', p. 654.
78. Megaw, p. 14.
79. J. D. Graham, 'International Aspects of Disease with Special Reference to Quarantine', *Far Eastern Association of Tropical Medicine. Transactions of the Seventh Congress, British India 1927*, vol. 1 (Calcutta: Thacker's Press, 1928), p. 473. By 1937 Indianization of the Public Health Services was almost complete, and only a handful of public health posts were held by Europeans. British Social Hygiene Council, *Empire Social Hygiene Year-Book, 1937* (London: George Allen and Unwin, 1937), pp. 409–10.
80. Sugata Bose and Ayesha Jalal, *Modern South Asia: History, Culture, Political Economy* (New York: Routledge, 2004), pp. 104–5.
81. Bose and Jalal (2004), p. 84.

82. Ibid., pp. 104–5. David Washbrook, 'The Rhetoric of Democracy and Development in Late Colonial India', *Nationalism, Democracy, and Development*, Sugata Bose and Ayesha Jalal (eds) (Delhi: Oxford University Press, 1997), pp. 41–3.

83. Ranajit Guha, 'Dominance without Hegemony and Its Historiography', *Subaltern Studies VI*, Ranajit Guha (ed.) (Delhi: Oxford University, 1989).

84. David Cannadine, 'The Empire Strikes Back', *Past and Present* (1995): 188.

85. 'Summary Report of the Inaugural Meeting of the Central Advisory Board of Health. Held in Simla on 22nd and 23rd June 1937' (Simla: Government of India Press, 1937) 17, IOR/v/25/840/70.

86. *Proceedings of the Third All-India Sanitary Conference held at Lucknow in January 1914, and Resolution of the Government of India on Sanitation in India* (London: His Majesty's Stationary Office, 1914), Appendix 7.

87. GOI, Education (Sanitary), Jun. 1920, A. 35-47, NAI.

88. Samanta, pp. 188, 191. David Arnold, '"An Ancient Race Outworn": Malaria and Race in Colonial India, 1860–1930', *Race, Science and Medicine, 1700–1960*, Bernard Harris and Waltraud Ernst (eds) (London: Routledge, 1999), p. 128.

89. 'The Future of Malaria Control in India', *IMG*, 62 (1927): 28–33.

90. Schüffner, p. 507.

91. Russell, 'Malaria in India', pp. 658, 663.

92. G. Covell, 'A Note on the Method Used to Combat Rural Malaria in India', *Intergovernmental Conference of Far-Eastern Countries on Rural Hygiene: Preparatory Papers relating to British India* (Geneva: League of Nations, 1937), p. 81.

93. Sandeep Sinha, *Public Health Policy and the Indian Public: Bengal 1850–1920* (Calcutta: Vision Publications, 1998), pp. 138–9.

94. Leslie, 'An Address on Malaria in India', pp. 1483–4. Prisoners were used to conduct experiments on the efficacy of certain doses of quinine as a preventative and as a cure.

95. A. C. Chatterji, 'Note on Public Health Organisation in Bengal', *Intergovernmental Conference of Far-Eastern Countries on Rural Hygiene: Preparatory Papers relating to British India* (Geneva: League of Nations, 1937), pp. 161–2.

96. C. M. Ganapathy, 'Public Health Administration in Madras Presidency', *Intergovernmental Conference of Far-Eastern Countries on Rural Hygiene: Preparatory Papers relating to British India* (Geneva: League of Nations, 1937), pp. 281–4.

97. As early as 1914, the GOI had articulated its intention to maintain control over research while decentralizing other branches of sanitation, recognizing that while the 'general direction of a policy of public health must remain with the central Government, all detailed control and executive action are, and will be, left to local Governments'. *Proceedings of the Third All-India Sanitary Conference*, Appendix 4.

98. Russell, 'A Note on the Central Government's Health Organization', pp. 11–12.

99. LON, 8C Health, R5961, 27535/21313, 'Note on the Quinine Position in India', pp. 2–3. GOI, Education, Health and Lands (Agriculture) Aug. 1926, A. 14-16, NAI.

100. *Ninth Conference of Medical Research Workers held at Calcutta from 30th November to 5th December 1931* (Simla: Government of India Press, 1932), pp. 46–7.
101. Letter from C. C. Calder, Director of the Botanical Survey to R. Littlehailes, Educational Commissioner with the GOI, 4 Jan 1932, in GOI, Education, Health and Lands (Education), 14-8/32. 1932, NAI. GOI, Education, Health and Lands (Education), 22-1/33-E. 1933, NAI.
102. GOI, Education, Health and Lands (Education), 14-8/32. 1932, NAI.
103. 'The Malaria Policy', *IMG*, 66 (1931): 520.
104. S. R. Christophers and W. F. Harvey, 'Malaria Research and Preventive Measures Against Malaria in the Federated Malay States and in the Dutch East Indies', *Indian Journal of Medical Research*, 10 (1922–23): 771.
105. 'Annual Report' (1926): 218–19. The Epidemic Diseases Act of 1897 was not applicable because malaria was not considered a dangerous epidemic disease like plague or cholera, and the act was intended only for temporary measures. GOI, Education, Health and Lands (Health), 44-8/34-H. 1934, NAI.
106. GOI, Education (Sanitary), Feb. 1918, IOR/p/10364.
107. G. Covell, 'The Malaria Survey of India, 1927–1937', *The Journal of the Malaria Institute of India* (Mar. 1938): 2–5, 11.
108. *Report of the Malaria Commission on its Study Tour in India* (Geneva: League of Nations, 1930), p. 74. Since ports and embankments caused by railroad construction were often ripe breeding grounds for mosquitoes, port trusts and railway companies were at times involved in malaria control measures.
109. N. Gangulee, *Health and Nutrition in India* (London: Faber and Faber, 1939), p. 149.
110. V. Shiva Ram and Brij Mohan Sharma, *India and the League of Nations* (Lucknow: Upper India Publishing House, 1932), pp. 136–8, 167.
111. D. N. Verma, *India and the League of Nations* (Patna: Bharati Bhawan, 1968), pp. 213–4. The GOI invited the League's commission to tour India in 1929. The Commission advised health administrations on malaria control, collected information on malaria epidemiology, and promoted international and national malaria research.
112. LON, 8C Health, R5953, 23526/1911, Letter from Ciuca to Christophers, 20 Oct. 1930.
113. LON, 8C Health, R6165, 911/911, Letter from Rajchman to Graham, 20 Jan. 1933. LON, 8C Health, R6166, 14000/911, Letter from Rajchman to Russell, 28 Dec. 1934.
114. LON, 8C Health, R5953, 25128/1411. Tellingly, Ross, who advocated mosquito eradication, was not invited to participate on the League's commission. Socrates Litsios, 'Malaria and International Health Organizations (A brief overview of the 20th century)', www.rockarch.org/publications/conferences (accessed 15 November 2008).
115. *Indian Social Reformer* (3 Aug. 1929) in LON, 8C Health, R5949, 503/503. GOI, Education, Health and Lands (Health) Aug. 1929, B. 179, NAI.
116. Gordon Harrison, *Mosquitoes, Malaria and Man: A History of the Hostilities Since 1880* (New York: E.P. Dutton, 1978), p. 183.
117. *Report of the Malaria Commission on its Study Tour in India*, pp. 73–4.
118. Ram and Sharma, p. 111.

119. 'Note on the Quinine Policy of India by the Malaria Survey of India', *Twelfth Conference of Medical Research Workers held at Calcutta from 26th November to 1st December 1934* (Simla: Government of India Press, 1935), pp. 144–7.
120. King, 'The Prevention of Malaria in India'.
121. Rai Bahadur and G. C. Chatterji, 'Biological Controls of Malaria in the Rural Areas of Lower Bengal', *Modern Review*, 54.6 (1933): 647.
122. Sinton, *What Malaria Costs India*, pp. 74–5.
123. Bahadur and Chatterji, p. 647.
124. Provincial governments did selectively use chemical methods of mosquito control in targeted areas. Covell, 'A Note on the Method Used to Combat Rural Malaria in India', pp. 73–5.
125. 'Note on the Quinine Policy of India by the Malaria Survey of India', pp. 144–9. Synthetic anti-malaria drugs came into therapeutic use in 1927, but the Malaria Survey advocated using cinchona alkaloids over the newer synthetic drugs, plasmoquine and atebrin, because the alkaloids were just as effective, did not require medical supervision to monitor toxicity, and were as cheap or cheaper to produce. 'The Therapeutics of Malaria', *Quarterly Bulletin of the Health Organisation of the League of Nations*, 2 (1933): 275.
126. *Report of the Intergovernmental Conference of Far-Eastern Countries on Rural Hygiene. Held at Bandoeng (Java), August 3–13, 1937* (Geneva: League of Nations, 1937), p. 89.
127. 'Summary Report of the Inaugural Meeting of the Central Advisory Board of Health', p. 25.
128. LON, 8C Health, R6163, 1242/1136, Letter from Graham to Rajchman, 11 Mar. 1933.
129. Dietmar Rothermund, *India in the Great Depression, 1929–1939* (New Delhi: Manohar, 1992), p. 5, Chapters 3 and 4. Tomlinson, p. 134. Bose, *Peasant Labour and Colonial Capital*, p. 184. Sugata Bose, *A Hundred Horizons: the Indian Ocean in the Age of Global Empire* (Cambridge: Harvard University Press, 2006), p. 214.
130. Strickland, p. 31.
131. Letter from Graham to Rajchman.
132. Russell, 'Malaria in India', pp. 663–4.
133. LON, 8B Health, R6157, 3031/378. The Malaria Survey of India maintained a hatchery of larvivorous fish for use by malaria workers. Covell, 'A Note on the Method Used to Combat Rural Malaria in India', p. 71.
134. Bahadur and Chatterji, pp. 648–9.
135. From 1938, the GOI did begin experimenting with insecticides to kill mosquitoes. Harrison, *Mosquitoes, Malaria and Man*, p. 211.
136. Quinine was produced from *Cinchona ledgeriana*, whereas totaquina was obtained from *Cinchona succirubra* or *Cinchona robusta*. LON, 8CHealth, R6174, 34241/1561, Letter from Wilson to Ciuca, 15 Aug. 1939. Cf. Sunil S. Amrith, *Decolonizing International Health: India and Southeast Asia, 1930–65* (Houndmills, Basingstoke: Palgrave Macmillan, 2006).
137. 'The Therapeutics of Malaria', p. 277. *The Treatment of Malaria*, p. 125.
138. 'Annual Report of the Public Health Commissioner with the Government of India for 1937, vol. 1', p. 775.
139. Hehir, pp. 289–90.
140. LON, 8C Health, R6164, 903/903. LON, 8C Health, R6173, 34241, 1561.

141. LON, 8C Health, R6164, 903/903. LON, 8C Health, R6173, 34241, 1561.
142. Strickland, pp. 3, 16–19, 28. Rural incomes had fallen catastrophically in these years, which may have prompted the GOI to lower prices.
143. GOI, Education (Sanitary), Jun. 1920, IOR/p/10830.
144. Hehir, pp. 293–7.
145. *Ninth Conference of Medical Research Workers*, pp. 45–6.
146. 'Cinchona Policy', *IMG*, 67 (1932): 392–3.
147. *IMG*, 73 (1938): 354–5.
148. GOI, Education, Health and Lands (Agriculture), May 1926, A. 1-34, NAI. The GOI allowed purchase of foreign quinine only if government supplies were inadequate to meet provincial needs.
149. *IMG*, 73 (1938): 354–5.
150. 'Annual Report' (1926): 215.
151. C. A. Gill 'The Theory and Practice of Malaria "Control"', *Far Eastern Association of Tropical Medicine: Transactions of the Seventh Congress, British India 1927*, vol. 2 (Calcutta: Thacker's Press, 1928), p. 634.
152. 'Annual Report' (1926): 210. 'The Therapeutics of Malaria', p. 278.
153. Hehir, p. 435.
154. 'Malnutrition and Malaria', *Modern Review*, 30.2 (1921): 266.
155. LON, C.H./Malaria/257, 21 Mar. 1938: 4–5.
156. Hehir, p. 435.
157. 'Annual Report' (1934): 3. The systematic and scientific study of malnutrition in British colonies began after World War I as a result of a link being made between diet and economic productivity. David Arnold, 'The 'discovery' of malnutrition and diet in colonial India', *The Indian Economic and Social History Review*, 31.1 (1994): 2, 19.
158. Assigning blame for deficient diets to native backwardness was a convenient excuse for officials to ignore the more recent interwar origins of malnutrition. Michael Worboys, 'The Discovery of Colonial Malnutrition between the Wars', *Imperial Medicine and Indigenous Societies*, David Arnold (ed.) (Manchester: Manchester University Press, 1988), pp. 222–3.
159. Hehir, p. 246.
160. Schüffner, p. 508.
161. LON, C.H./Malaria/135 20 in 8C Health, R5949, 3858/477.
162. 'Abstract of the Annual Report for 1935 of the Public Health Commissioner with the Government of India', *IMG*, 73 (1938): 248.
163. Harrison, *Mosquitoes, Malaria and Man*, pp. 184–6.
164. 'Annual Report of the Public Health Commissioner with the Government of India for 1936', *IMG*, 73 (1938): 708. Before the GOI began differentiating malaria from fever mortality, it assumed that 25 percent of fever deaths were attributable to malaria. This estimate was based on special enquiries, dispensary reports, and other official sources. Leslie, 'An Address on Malaria in India', p. 1483. J. T. W. Leslie, 'Malaria in India', *Proceedings of the Imperial Malaria Conference held at Simla in October 1909* (Simla: Government Central Branch Press. 1910).
165. Randall M. Packard, *The Making of a Tropical Disease: A Short History of Malaria* (Baltimore: Johns Hopkins University Press, 2007), p. 115.
166. Mark Harrison, '"Hot Beds of Disease"; Malaria and Civilization in Nineteenth-Century British India', *Parassitologia*, 40 (1998): 11.

167. Samanta, pp. 140–1. Raymond E. Dumett, 'The Campaign against Malaria and the Expansion of Scientific Medical and Sanitary Services in British West Africa, 1898–1910' *African Historical Studies*, 1.2 (1968): 165. Ironically, Manson stressed an approach that privileged further research over practical action and individual hygienic precautions, such as bed nets and regular quinine use. The latter approach was fine for Europeans, but as Ross argued, impracticable for the masses.
168. Helen J. Power, *Tropical Medicine in the Twentieth Century: A History of the Liverpool School of Tropical Medicine, 1898–1990* (London: Kegan Paul, 1999), p. 15.
169. Sinton, *What Malaria Costs India*, p. 79.
170. Ibid., p. 108.
171. 'Annual Report' (1926): 212.
172. *Health Organisation in British India*, p. 37.
173. Anil Kumar, *Medicine and the Raj: British Medical Policy in India, 1835–1911* (Walnut Creek: Altamira Press, 1998), p. 187.
174. Anil Kumar, 'The Indian Drug Industry under the Raj, 1860–1920', *Health, Medicine and Empire: Perspectives on Colonial India*, Biswamoy Pati and Mark Harrison (eds) (New Delhi: Orient Longman, 2001), p. 369.
175. Packard, p. 122. Harrison, *Mosquitoes, Malaria and Man*, p. 173. The malaria parasite can exist in both sexual and asexual forms in the human host's blood. Adequate treatment requires killing both forms.
176. *Health Organisation in British India*, p. 38.
177. Hehir, pp. 427–8.
178. Amrith, *Decolonizing International Health*, p. 40. Cf. *Report of the Intergovernmental Conference of Far-Eastern Countries on Rural Hygiene*.
179. V. R. Muraleedharan, 'Malady in Madras: The Colonial Government's Response to Malaria in the Early Twentieth Century', *Science and Empire: Essays in India Context (1700–1947)*, Deepak Kumar (ed.) (Delhi: Anamika Prakashan, 1991), p. 111.
180. Tomlinson, pp. 27, 105. Rothermund, pp. 5–9. G. Balachandran, 'Introduction', *India and the World Economy, 1850–1950*, G. Balanchandran (ed.) (New Delhi: Oxford University Press, 2003), p. 9. P. J. Cain and A. G. Hopkins, *British Imperialism, 1688–2000* (Harlow: Longman, 2002), p. 545. Home charges included the costs of Indian government in Britain, pensions to returned military and civil servants, and guaranteed interest on rail construction. John Adams and Robert Craig West, 'Money, Prices, and Economic Development in India, 1861–1895', *Journal of Economic History*, 39.1 (1979): 59.

## 4 From Panama to Khartoum – Yellow Fever Inches Closer to Home

1. Yellow fever virus has never been recorded in India or in other areas of Asia, for reasons that have more to do with biology than with the Government of India's policies then or now. Demographic and biological factors offer possible explanations: (1) Yellow fever occurs in remote areas and affects mostly individuals working in subsistence farming who rarely travel internationally. Thus

they are unlikely to introduce the disease into an uninfected country. (2) Cross-protection could be provided by dengue fever immunity. (3) *Aedes aegypti* strains in India have low vector competence for the yellow fever virus. Thomas P. Monath, 'Yellow Fever', *Vaccines*, 3rd edn, Stanley A. Plotkin and Walter A. Orenstein (eds) (Philadelphia: W.B. Saunders Company, 1999), p. 830.

2. G. Covell, 'Foreword' in R. K. Mhatre, *A Survey of Aedes Mosquitoes in Bombay and the Measures Suggested for Their Control* (Bombay: British India Press, 1934). At the time, public health experts thought that even if people living in the East Indies had some level of cross-immunity conferred by dengue fever infection, this did not preclude the possibility of a yellow fever catastrophe, although it might have mitigated its effect. E. P. Snijders, 'The Yellow Fever Problem in the Far East', in *Far Eastern Association of Tropical Medicine: Transactions of the Eighth Congress held in Siam 1930*, vol. 1 (Bangkok: Bangkok Times Press, 1931), p. 140.

3. Yellow fever is caused by a flavivirus, which is transmitted to humans and to non-human primates by the *Stegomyia fasciata* (later renamed *Aedes aegypti*) mosquito. The incubation period in humans is three to six days. The blood of patients is infective to mosquitoes shortly before the onset of fever and during the first three to five days of illness. The extrinsic incubation period in *Aedes aegypti* is 9–12 days at typical tropical temperatures. Once the mosquito is infected with the virus, it remains so for the rest of its life. Yellow fever is highly communicable where there are susceptible people and an abundance of the mosquito vector. David L. Heymann (ed.), *Control of Communicable Diseases Manual* (Washington: American Public Health Association, 2004), pp. 595–7.

4. *Conférence Sanitaire Internationale de Paris. 10 Octobre–3 Decembre 1903. Procès-Verbaux.* (Paris: Imprimerie Nationale, 1904), p. 351.

5. The Government of Bengal told the Government of India that its port authorities did not have the legal power to implement anti-Stegomyia operations because the city's sanitary organization came under the jurisdiction of several independent health authorities. Since Calcutta was not yet threatened by an outbreak, the Government of Bengal could not invoke the Epidemic Diseases Act. The Government of India understood the legal difficulties but suggested that while Bengal was sorting this out that it at least improve sanitation for certain ports. GOI, Education (Sanitary), Sept., Oct. 1916, IOR/p/9946.

6. R. K. Mhatre, *A Survey of Aedes Mosquitoes in Bombay and the Measures Suggested for Their Control* (Bombay: British India Press, 1934), pp. 1–2.

7. Health officials in India had earlier proposed a variation of this idea when they had called for better sanitary controls in the Hedjaz to prevent pilgrims from contracting plague or cholera and bringing it back to India.

8. GOBo, General (General), Feb. 1911, IOR/p/ 8829. The LGB would become the British Ministry of Health in 1919.

9. 'Proceedings of the Imperial Malaria Committee held in Bombay on 16th and 17th November 1911', *Paludism, Being the Transactions of the Committee for the Study of Malaria in India*, 4 (1912): 2.

10. Mhatre, p. 2.

11. *The Proceedings of the Second All-India Sanitary Conference held at Madras in November 1912*, vol. 1 (Simla: Government Central Branch Press, 1913),

pp. 7–8. 'Proceedings of the Imperial Malaria Committee held in Bombay on 16th and 17th November 1911', *Paludism, Being the Transactions of the Committee for the Study of Malaria in India*, 4 (1912): 2. GOI, Education (Sanitary), Jun. 1912, IOR/p/8947.

12. S. P. James, 'The Protection of India from Yellow Fever', p. 3.
13. Ibid., p. 5.
14. GOI, Education (Sanitary), Feb. 1913, IOR/p/9199.
15. Frank G. Clemow, *The Geography of Disease* (Cambridge: University Press, 1903), pp. 526–7.
16. The 1912 convention was the first to address yellow fever in significant detail, elevating its profile as a major infectious disease requiring international cooperation to control.
17. *Conférence Sanitaire Internationale de Paris. 7 Novembre 1911–17 Janvier 1912. Procès-Verbaux* (Paris: Imprimerie Nationale, 1912), pp. 168–9, 392. Britain's stance toward India on this matter seems at odds with the general trend from the 1870s to the early 1910s of the India Office in London exercising increasing authority over long-term planning and implementation of policy. Arnold P. Kaminsky, *The India Office, 1880–1910* (New York: Greenwood Press, 1986), p. 151.
18. *Procès-Verbaux*, 1911–12, pp. 168–9, 392.
19. Mhatre, pp. 4–5.
20. GOI, Education (Sanitary), Aug. 1913, IOR/p/9200. GOI, Education (Sanitary), Mar. 1914, IOR/p/9448. GOI, Education (Sanitary), Oct. 1916, IOR/p/9946.
21. 'Annual Report' (1917): 139. GOI, Education (Sanitary) Feb. 1918, B. 96, NAI.
22. GOI, Education (Sanitary), Nov. 1917, IOR/p/10169.
23. Ibid.
24. Ibid.
25. GOI, Education (Sanitary), Aug. 1917, IOR/p/10169.
26. GOI, Education (Sanitary), Jan. 1916, IOR/p/9946. GOI, Education (Sanitary), Feb. 1916, IOR/p/10169. GOI, Education (Sanitary), Feb. 1920, NAI.
27. GOI, Education (Sanitary), Sept., Oct. 1916, IOR/p/9946.
28. GOBo, General (General), Mar. 1919, IOR/p/10538.
29. The Government of India's public health expenditures as a percent of total government expenditure was a meager 0.37 percent in 1920–1. *East India: Accounts and Estimates, 1921–1922* (London: Eason and Son, 1921).
30. GOI, Education (Sanitary) Apr 1921, A. 1–6, NAI.
31. GOI, Education (Sanitary) Jun. 1919, B. 89–96, NAI.
32. GOI, Education and Health (Sanitary), May 1922, A. 25–43, NAI.
33. GOI, Education (Sanitary), Feb. 1920, IOR/p/10830.
34. GOI, Education (Sanitary), Jul. 1920, IOR/p/10830.
35. GOI, Education (Sanitary), Feb. 1920, IOR/p/10830.
36. GOI, Education (Sanitary), Jun. 1918, IOR/p/10364. GOI, Education (Sanitary) Nov. 1917, A. 1-2, NAI.
37. GOI, Education (Sanitary), Feb. 1920, IOR/p/10830.
38. Mhatre, pp. 4–5.
39. F. Norman White, *The Prevalence of Epidemic Disease and Port Health Organisation and Procedure in the Far East* (Geneva: League of Nations, 1923), p. 147.

40. Ibid., p. 148. The exact incubation period had not yet been determined for yellow fever, so for purposes of international health regulations, 18 days was used as the minimum monitoring and detection period following the last death, recovery, or isolation of a yellow fever case. GOI, Education (Sanitary), Jul. 1920, A. 17–19, NAI.

41. GOI, Education (Sanitary), Jul. 1920, IOR/p/10830.

42. *Proceedings of the Third All-India Sanitary Conference held at Lucknow in January 1914, and Resolution of the Government of India on Sanitation in India* (London: His Majesty's Stationary Office, 1914) 16. GOI, Education (Sanitary), Mar. 1913, IOR/p/9199.

43. GOI, Education and Health (Sanitary), May 1922, IOR/p/11211.

44. Lorna Weir and Eric Mykhalovskiy, *Global Public Health Vigilance: Creating a World on Alert* (New York: Routledge, 2010), pp. 3, 7–9.

45. GOI, Education (Sanitary), Jul. 1920, IOR/p/10830.

46. GOI, Education and Health (Sanitary), May 1922, A. 25-43, NAI.

47. 'The Need for a Public Health for India', *Indian Medical Gazette*, 62 (1927): 579–80.

48. A. R. Wellington, *Hygiene and Public Health in India. Report on Conditions Met with During the Tour of the League of Nations Interchange of Health Officers* (Kuala Lumpur: Federated Malay States Government Press, 1929), p. 28.

49. J. D. Graham, 'International Aspects of Disease with Special Reference to Quarantine', *Far Eastern Association of Tropical Medicine. Transactions of the Seventh Congress, British India 1927*, vol. 1 (Calcutta: Thacker's Press, 1928), p. 470.

50. GOB, Local (Public Health), Oct. 1935, IOR/p/12054. By the end of 1931, the Eastern Bureau in Singapore was collecting epidemiological intelligence from 153 ports in 50 countries in Asia, Africa, and Oceania and distributing information on outbreaks to public health administrations. By 1935, 185 regional ports were sending information to Singapore. 'Report of the Health Organisation', *Quarterly Bulletin of the Health Organisation* (Sept. 1932): 117. Alison Bashford, 'Global Biopolitics and the History of World Health', *History of the Human Sciences*, 19.1 (2006): 72.

51. GOI, Education (Sanitary), Jul. 1920, IOR/p/10830. GOBo, General (General), Jan. 1920, IOR/p/10783. Yellow fever was widespread throughout the Congo basin and Central Africa. Officials thought that the British north-south transcontinental rail could facilitate disease spread to the East African seaboard, but the railway was never completed. 'Annual Report' (1928).

52. P. J. Barraud, 'The Distribution of "Stegomyia Fasciata" in India with Remarks on Dengue and Yellow Fever', *Indian Journal of Medical Research*, 16 (1928–29): 380.

53. Mhatre, pp. 6–8.

54. Snijders, p. 133.

55. J. A. Sinton, 'Suggestions with regard to the prevention of the spread of Yellow Fever to India by Air Traffic, with special reference to Insect Transmission', *Health Bulletin*, 20 (1934): 9.

56. 'Is Public Health Worth While'? *IMG*, 65 (1930): 708. 'Annual Report' (1929): 150.

57. Heather Bell, *Frontiers of Medicine in the Anglo-Egyptian Sudan, 1899–1940* (Oxford: Clarendon Press, 1999), p. 171.

58. Snijders, pp. 145, 149. 'Aerodrome' was the common term for airfield before World War II.

59. 'The Cape Town Conference', *Quarterly Bulletin of the Health Organisation of the League of Nations*, 2 (1933): 22. Making an aerodrome anti-amaryl required the elimination of all mosquitoes in the vicinity and provision of mosquito-proof accommodation for travelers in transit. The nearest human dwelling had to be more than two kilometers from the airport, which was the flight range of the *Aedes aegypti* mosquito. Bell, pp. 175–6.

60. 'Annual Report' (1932): 177. Advances in etiological knowledge had narrowed the internationally accepted incubation time to three to six days. Scientists discovered that the mosquito vector was an urban species that had to survive at least 12 days after infection before it could transmit the virus, and infected mosquitoes rarely traveled more than 100 meters. S. P. James, 'Connaissances Récemment Acquises sur la Fièvre Jaune', *Bulletin Mensuel de l'Office International D'Hygiène Publique*, 25 (1933): 62–3.

61. LON, 8A Health, R6080, 40421/2652, Letter from Russell to Rajchman, 13 Dec. 1938. LON, 8A Health, R6062, 9361/985, Letter from Russell to Rajchman, 17 Dec. 1934.

62. LON, 8A Health, R6080, 2964/2652, Letter from Biraud to Mhatre, 19 Nov. 1935. George S. Buchanan, 'Letters to the Editor: Yellow Fever: Transmission by Aircraft', *Times* (25 Feb. 1933).

63. 'Yellow Fever: Transmission over Continents', *Times* (24 Feb. 1933).

64. C. A. Sprawson, 'The Position of India in Regard to the Yellow Fever Question', *Quarterly Bulletin of the Health Organisation*, 5 (1936), pp. 87–8. Mouse protection tests assessed the presence of antibodies in the sera of mice to see if yellow fever was present in the area.

65. 'Annual Report of the Public Health Commissioner for the Year 1937', *IMG*, 73 (1938): 251.

66. Bell, p. 169.

67. 'Annual Report' (1934): 273–4.

68. James, 'Connaissances Récemment Acquises sur la Fièvre Jaune', pp. 62–3.

69. 'Report on the Session of the Office International D'Hygiène Publique, held in Paris from 25th April to 4th May, 1932', *IMG*, 68 (1933): 119–20.

70. 'Annual Report' (1937): 251. Disinsectization was the contemporary term for ridding airports, airplanes, and ships of disease-carrying insects. Previously, the Epidemic Diseases Act had been applied only to diseases actually epidemic in India. It was not originally intended to prevent the introduction of a disease.

71. Sprawson, pp. 87–8. 'The Yellow Fever Peril', *IMG*, 74 (1939): 699.

72. Snijders, p. 140.

73. 'The Cape Town Conference', p. 19.

74. 'Annual Report' (1937): 251.

75. 'Annual Report' (1935): 101–2. Quote from Sprawson at the 1935 Pan-African Health Conference.

76. 'The Rockefeller Foundation, New York: Annual Report, 1936', *IMG*, 73 (1938): 184.

77. Sudan acted as a critical barrier to the extension of infection to Egypt, Iraq, and India. The Government of Sudan made arrangements with West African governments, so that no passenger could leave West Africa for Khartoum

and Egypt unless the passenger had resided in a yellow-fever-free area for a minimum of six days prior to embarkation. 'Annual Report' (1937): 227.

78. 'Annual Report' (1937): 251. Two species of monkeys found throughout almost the whole of India had demonstrated susceptibility to yellow fever. Barraud, p. 379.

79. Sinton, 'Suggestions with regard to the prevention of the spread of Yellow Fever to India by Air Traffic', pp. 2–6. Some officials viewed the religious protection of monkeys in certain parts of India as an additional obstacle to disease control.

80. Snijders, pp. 133–8.

81. Peter Bennett', Understanding Responses to Risk: Some Basic Findings', *Risk Communication and Public Health*, Peter Bennett and Sir Kenneth Calman (eds) (Oxford: Oxford University Press, 1999), p. 8.

82. Andrew Lakoff, 'Introduction', *Disaster and the Politics of Intervention*, Andrew Lakoff (ed.) (New York: Columbia University Press, 2010), p. 4.

83. Filippa Lentzos, 'The Pre-History of Biosecurity: Strategies of Managing Risks to Collective Health', *Biosecurity: Origins, Transformations and Practices*, Brian Rappeprt and Chandré Gould (eds) (Houndmills, Basingstoke: Palgrave Macmillan, 2009), pp. 31, 36. Preparedness is an approach to dealing with a hazard that cannot be statistically calculated but is potentially cata-strophic. Lyle Fearnley, 'Redesigning Syndromic Surveillance for Biosecurity', *Biosecurity Interventions: Global Health and Security in Question*, Andrew Lakoff and Stephen J. Collier (eds) (New York: Columbia University Press, 2008), p. 64.

84. Frédéric Keck, 'From Mad Cow Disease to Bird Flu: Transformations of Food Safety in France', *Biosecurity Interventions: Global Health and Security in Question*, Andrew Lakoff and Stephen J. Collier (eds) (New York: Columbia University Press, 2008), pp. 211, 213. Poul Harremoës et al., 'Introduction', *The Precautionary Principle in the 20th Century: Late Lessons from Early Warnings*, Poul Harremoës et al. (eds) (London: Earthscan, 2002), pp. 4–7.

85. Bell, p. 166. Cf. Keith Wailoo, *Drawing Blood: Technology and Disease Identity in Twentieth-Century America* (Baltimore: The Johns Hopkins University Press, 1997).

86. GOB, Local (Public Health), Nov. 1936, IOR/p/12080.

87. GOI, Education (Sanitary – Deposit) Mar 1920, no 19, NAI.

88. Jenifer Van Vleck, 'The Logic of the Air: Visions of Aviation and Empire, 1938–1945', paper presented at Harvard Graduate Student Conference on International History, Harvard University, Cambridge, MA, 16 March 2007. The Government of India instituted other precautions. It maintained emergency stocks of the yellow fever vaccine, courtesy of the Rockefeller Foundation, at the Haffkine Institute in Bombay. Public health and research officers were encouraged to obtain inoculation training at the Wellcome Laboratory when in England. Since most public health officials in India had little experience with diagnosing the disease, the Government of India made an expert diagnosis unit available to provincial and state governments for use in any suspected cases, and the Malaria Institute of India placed an expert mosquito control unit at the disposal of provincial governments should yellow fever break out. 'Annual Report' (1940): 1–2.

89. 'The Yellow Fever Peril', pp. 698–9.

90. 'Report of the Health Organisation', p. 118. Snijders, pp. 146–7. GOI, Education (Health), May 1931, IOR/p/11933. GOB, Local (Public Health), Dec. 1936, IOR/p/12080. Although the Government of India was working to meet the standards of the 1926 international sanitary convention in terms of port sanitation and the regulation of international shipping, it still had not ratified the convention as of March 1939, possibly because constitutionally, the Government of India could not force provinces to improve the sanitary conditions of ports.

91. Norman Howard-Jones, *The Scientific Background of the International Sanitary Conferences, 1851–1938* (Geneva: World Health Organization, 1975), pp. 90, 93.

92. Mhatre, p. 6.

# 5  Disease as Prism

1. Thomas R. Metcalf, *Forging the Raj: Essays on British India in the Heyday of Empire* (New Delhi: Oxford University Press, 2005), pp. 282–4. Frederick Cooper, *Colonialism in Question: Theory, Knowledge, History* (Berkeley: University of California Press, 2005), p. 28. Ballantyne, *Orientalism*, p. 15. David Armitage, 'Three Concepts of Atlantic History', *The British Atlantic World, 1500–1800*, David Armitage and Michael J. Braddick (eds) (Houndmills, Basingstoke: Palgrave Macmillan, 2002). Kathleen Wilson, *A New Imperial History: Culture, Identity and Modernity in Britain and the Empire, 1660–1840* (Cambridge: Cambridge University Press, 2004).

2. After World War I, resurgent nationalism coincided with the decline of the subimperial system as tariff barriers and immigration restrictions due to economic crises arose. This led India to become increasingly enveloped in notions of self-sufficient development from the 1920s. Thomas R. Metcalf, *Imperial Connections: India in the Indian Ocean Arena, 1860-1920* (Berkeley: University of California Press, 2007), pp. 11–13.

3. Tony Ballantyne, 'Empire, Knowledge and Culture: From Proto-Globalization to Modern Globalization', *Globalization in World History*, A. G. Hopkins (ed.) (New York: W.W. Norton & Company, 2002). Tony Ballantyne, *Orientalism and Race: Aryanism in the British Empire* (London: Palgrave Macmillan, 2002), pp. 3, 15.

4. Douglas Haynes, *Imperial Medicine: Patrick Manson and the Conquest of Tropical Disease* (Philadelphia: The University of Pennsylvania Press, 2001), p. 48.

5. Richard H. Grove, *Green Imperialism: Colonial, Expansion, Tropical Island Edens, and the Origins of Environmentalism, 1600–1860* (Cambridge: Cambridge University Press, 1995).

6. Cf. I. J. Catanach, 'Plague and the Tensions of Empire: India 1896–1918', *Imperial Medicine and Indigenous Societies*, David Arnold (ed.) (Manchester: Manchester University Press, 1988).

7. Cf. Dorothy Porter, *Health, Civilization and the State: A History of Public Health from Ancient to Modern Times* (London: Routledge, 1999).

8. Neville M. Goodman, *International Health Organizations and Their Work* (Philadelphia: The Blakiston Company, 1952), pp. 101–2. J. C. Coyajee, *India and the League of Nations* (Madras: Waltair, 1932), pp. 165–8.

9. 'The Suppression of Plague and Malaria in India', *IMG*, 46 (1911): 429–30.
10. Sunil S. Amrith, *Decolonizing International Health: India and Southeast Asia, 1930–65* (Houndsmill: Palgrave Macmillan, 2006), pp. 7–8.
11. P. K. Manning, 'Managing Risk: Managing Uncertainty in the British Nuclear Installations Inspectorate', *Law and Policy*, 11.3 (1989): 351.
12. Cf. Lee Clarke and James F. Short, Jr. 'Social Organization and Risk: Some Current Controversies', *Annual Review of Sociology*, 19 (1993): 375–99.
13. Niklas Luhmann, *Risk: A Sociological Theory* (New Brunswick: Aldine Transaction, 2002), pp. 2–3.
14. Norman Howard-Jones, *The Scientific Background of the International Sanitary Conferences, 1851–1938* (Geneva: World Health Organization, 1975), p. 76.
15. See Peter Baldwin, *Contagion and the State in Europe, 1830–1930* (Cambridge: Cambridge University Press, 1999) for his discussion of geoepidemiological trajectories.
16. Ibid., pp. 211–14, 226.
17. Claire Hooker, 'Sanitary Failure and Risk: Pasteurisation, Immunisation and the Logics of Prevention', *Contagion: Epidemics, History and Culture from Smallpox to Anthrax*, Alison Bashford and Claire Hooker (eds) (Annandale: Pluto Press Australia, 2002 [2001]), pp. 144–5.
18. Valeska Huber, 'The Unification of the Globe by Disease? The International Sanitary Conferences on Cholera, 1851–1894', *Historical Journal*, 49.2 (2006): 467.
19. David P. Fidler, 'The Globalization of Public Health: The First 100 Years of International Health Diplomacy', *World Health Organization*, 79 (2001): 847.
20. Cf. Richard Tuck, *The Rights of War and Peace: Political Thought and the International Order from Grotius to Kant* (Oxford: Oxford University Press, 1999).
21. Edward Hallett Carr, *The Twenty Years' Crisis, 1919–1939* (London: Macmillan & Co., 1954), p. 97. Hedley Bull, *The Anarchical Society: A Study of Order in World Politics* (New York: Columbia University Press, 1977), p. 139.
22. Alexandra Minna Stern and Howard Markel, 'International Efforts to Control Infectious Diseases, 1851 to the Present', *JAMA*, 292 (2004): 1474, 1478.
23. Louis Henkin argues that nations have a tendency to comply. Only if 'violation promises an important balance of advantage over cost' will nations disregard or violate international obligations. Louis Henkin, *How Nations Behave: Law and Foreign Policy* (New York: Columbia University Press, 1979), pp. 49–50, 58.
24. Obijiofor Aginam, 'International Law and Communicable Diseases', *Bulletin of the World Health Organization* (2002): 949.
25. 'Report of the Indian Delegation to the Secretary of State for India', 'Annual Report' (1926): 80.
26. J. D. Graham, 'Hygiene and Public Health', *IMG. Supplement – The Indian Medical Year, 1926. A Review* (1927): 16.
27. GOI, Education, Health and Lands (Health), Jun. 1926, IOR/p/11556.
28. 'India and the League of Nations', *IMG*, 63 (1928): 160. Quote from lecture by J. D. Graham to the Rotary Club, Calcutta on 6 Dec. 1927'.
29. Alison Bashford, '"The Age of Universal Contagion": History, Disease and Globalization', *Medicine at the Border: Disease, Globalization and Security, 1850 to the Present*, Alison Bashford (ed.) (New York: Palgrave Macmillan, 2006), pp. 5–6.
30. 'Annual Report' (1922): 108.

31. Cf. Sugata Bose, 'Instruments and Idioms of Colonial and National Development: India's Historical Experience in Comparative Perspective', *International Development and the Social Sciences: Essays on the History and Politics of Knowledge*, Frederick Cooper and Randall Packard (eds) (Berkeley: University of California Press, 1997).

32. Dean T. Jamison, 'Cost-Effectiveness Analysis: Concepts and Applications', *Oxford Textbook of Public Health*, Roger Detels et al. (eds), vol. 2 (Oxford: Oxford University Press, 2002), p. 906.

33. P. J. Cain and A. G. Hopkins, *British Imperialism, 1688–2000* (Harlow: Longman, 2002), p. 278.

34. Mark Harrison, 'Public Health and Medical Research in India, c. 1860–1914', diss, Oxford University, 1991, p. 270.

35. 'The Far Eastern Association of Tropical Medicine', *IMG*, 62 (1927): 280.

36. Arabinda Samanta, *Malarial Fever in Colonial Bengal, 1820–1939: Social History of an Epidemic* (Kolkata: Firma KLM, 2002), p. 204.

37. Dipesh Chakrabarty, *Habitations of Modernity: Essays in the Wake of Subaltern Studies* (Chicago: The University of Chicago Press, 2002), pp. 76–7.

38. Huber, pp. 471, 474.

39. Mark Harrison, *Public Health in British India: Anglo-Indian Preventive Medicine, 1859–1914* (Cambridge: Cambridge University Press, 1994), pp. 149, 200. Roger Jeffery, *The Politics of Health in India* (Berkeley: University of California Press, 1988), p. 101.

40. Frank M. Snowden, *The Conquest of Malaria: Italy, 1900–1962* (New Haven: Yale University Press, 2006), pp. 215–18.

41. 'Annual Report' (1926): 209.

42. Ibid.

43. John M. MacKenzie, 'Introduction', *Imperialism and the Natural World*, John M. MacKenzie (ed.) (Manchester: Manchester University Press, 1990), p. 12. Cf. Michael Worboys, 'The Imperial Institute: The State and the Development of the Natural Resources of the Colonial Empire, 1887–1923' and John M. MacKenzie, 'Experts and Amateurs: Tsetse, Nagana and Sleeping Sickness in East and Central Africa', *Imperialism and the Natural World*, John M. MacKenzie (ed.) (Manchester: Manchester University Press, 1990).

44. Nancy Stepan, 'The Interplay between Socio-Economic Factors and Medical Science: Yellow Fever Research, Cuba and the United States', *Social Studies of Science*, 8.4 (1978), p. 398.

45. 'Annual Report' (1926): 211.

46. 'Annual Report' (1927): 239.

47. 'The Need for a Public Health for India', *Indian Medical Gazette*, 62 (1927): 575.

48. David Washbrook, 'The Rhetoric of Democracy and Development in Late Colonial India', *Nationalism, Democracy, and Development*, Sugata Bose and Ayesha Jalal (eds) (Delhi: Oxford University Press, 1997), pp. 44–5.

49. V. R. Muraleedharan, 'Malady in Madras: The Colonial Government's Response to Malaria in the Early Twentieth Century', *Science and Empire: Essays in India Context (1700–1947)*, Deepak Kumar (ed.) (Delhi: Anamika Prakashan, 1991), pp. 110–11.

50. Baldwin, pp. 236–7.

51. 'Annual Report' (1926): 211.

52. Judith M. Brown, 'Imperial Facade: Some Constraints upon and Contradictions in the British Position in India, 1919–35', *Transactions of the Royal Historical Society* series 5, 26 (1976): 44, 51.
53. Deepak Kumar, 'Perceptions of Public Health: A Study in British India', *Maladies, Preventives and Curatives: Debates in Public Health in India*, Amiya Kumar Bagchi and Krishna Soman (eds) (Kolkata: Tulika Books, 2005), p. 49.
54. 'The Need for a Public Health for India', p. 582.
55. Ibid., p. 580.
56. British Social Hygiene Council, *Empire Social Hygiene Year-Book, 1937* (London: George Allen and Unwin, 1937), p. 408.
57. Sheila Jasanoff, 'Beyond Calculation: A Democratic Response to Risk', *Disaster and the Politics of Intervention*, Andrew Lakoff (ed.) (New York: Columbia University Press, 2010), p. 29.
58. Uday Singh Mehta, *Liberalism and Empire: A Study in Nineteenth-Century British Liberal Thought* (Chicago: The University of Chicago Press, 1999), pp. 190–1.
59. Thomas R. Metcalf, *Ideologies of the Raj* (Cambridge: Cambridge University Press, 1994), p. 43.
60. Cf. Ann Laura Stoler and Frederick Cooper, 'Between Metropole and Colony: Rethinking a Research Agenda', *Tensions of Empire: Colonial Cultures in a Bourgeois World*, Frederick Cooper and Ann Laura Stoler (eds) (Berkeley: University of California Press, 1997).
61. Partha Chatterjee, *The Nation and Its Fragments: Colonial and Postcolonial Histories* (Princeton: Princeton University Press, 1993).
62. Cf. David Arnold, *Colonizing the Body: State Medicine and Epidemic Disease in Nineteenth-Century India* (Berkeley: University of California Press, 1993).
63. Ira Klein, 'Medicine and Culture in British India', *Journal of Indian History*, 76–78 (1997–9): 122.
64. Ibid., p. 175.
65. 'Notification of Dangerous Diseases', *IMG*, 64 (1929): 211–13. 'Annual Report of the Public Health Commissioner with the Government of India for 1937, vol. 1', p. 776.
66. 'The Need for a Public Health for India', p. 575.
67. 'Annual Report' (1937): 135. A. J. H. Russell, 'Population and Public Health in India', *Far Eastern Association of Tropical Medicine. Transactions of the Seventh Congress, British India 1927*, vol. 1 (Calcutta: Thacker's Press, 1928), pp. 613–14, 633.
68. Sumit Guha, *Health and Population in South Asia: From Earliest Times to the Present* (London: Hurst, 2001), p. 158.
69. Chittabrata Palit, *Scientific Bengal: Science, Technology, Medicine and Environment under the Raj* (Delhi: Kalpaz Publications, 2006), p. 179.
70. Margaret Jones, *Health Policy in Britain's Model Colony: Ceylon (1900–1948)* (New Delhi: Orient Longman, 2004), p. 151.

## Epilogue: Swine Flu Redux

1. Andrew Martin and Clifford Krauss, 'Pork Industry Fights Concerns Over Swine Flu', *New York Times*, 28 April 2009, http://www.nytimes.com (accessed 6 November 2009).

2. 'H1N1 Flu: Monitoring the Nation's Response', United States Senate Committee on Homeland Security and Governmental Affairs, 21 October 2009.
3. Daniel Carty, 'Obama: Swine Flu "Not A Cause For Alarm"', *CBS News*, 27 April 2009, http://www.cbsnews.com/blogs/2009/04/27/politics/political-hotsheet/entry4970794.shtml (accessed 6 November 2009).
4. Jason Socrates Bardi, 'Grounding Flights Won't Stop Flu', *Live Science*, 1 October 2009, http://www.livescience.com (accessed 4 November 2009). Jon Cohen, 'Out of Mexico? Scientists Ponder Swine Flu's Origins', *Science* 324.5928, pp. 700–2. www.sciencemag.org (accessed 5 November 2009).
5. Nadim Audi, 'Culling Pigs in Flu Fight, Egypt Angers Herders and Dismays UN', *New York Times*, 30 April 2009, http://www.nytimes.com (accessed 4 November 2009). The Egyptian government did also issue an alert in May that pilgrims going on the *hajj* in December may be prevented from reentering Egypt to ensure that they did not infect others in the country. In response, Saudi Arabia's Grand Mufti Sheik Abdul-Aziz Al Sheikh stated that the possibility of exposure to swine flu during the *hajj* was exaggerated and accused pharmaceutical companies of spreading rumors to create panic and promote their drug sales. Sarah El Deeb, 'Egypt Warns of Post-Hajj Swine Flu Quarantine', *abc News*, 20 May 2009, http://abcnews.go.com/International/wireStory?id=7634743 (accessed 5 November 2009).
6. Daniel Woolls, 'Swine Flu Cases in Europe; Worldwide Travel Shaken', *Taiwan News*, 28 April 2009, http://www.etaiwannews.com (accessed 5 November 2009).
7. 'Cruise Ship Pacific Dawn Quarantined near Cairns over Swine Flu', *The Australian*, 27 May 2009, http://www.theaustralian.com.au (accessed 5 November 2009).
8. Brian Todd and Taylor Gandossy, 'China Quarantines US School Group over Flu Concerns', *CNN*, 28 May 2009, http://www.cnn.com (accessed 4 November 2009). US Department of State, Bureau of Consular Affairs, 'Travel Alert: China', 25 September 2009 http://travel.state.gov (accessed 4 November 2009). 'Tensions Escalate in Hong Kong's Swine-Flu Hotel', *Taipei Times*, 3 May 2009, http://www.taipeitimes.com (accessed 5 November 2009). Nicholas Kralev, 'KRALEV: US Warns Travelers of China's Flu Rules', *Washington Times*, 29 June 2009, http://washingtontimes.com (accessed 4 November 2009).
9. Sherry Towers and Zhilan Feng, 'Novel H1N1: Predicting the Course of a Pandemic', http://www.stat.purdue.edu/~stowers/flu/poster.pdf (accessed 20 November 2009).
10. 'HHS 2009 H1N1 Vaccine Development Activities', https://www.medical-countermeasures.gov/BARDA/MCM (accessed 13 December 2009).
11. Talea Miller, 'US Passes on Unlicensed H1N1 Vaccine Boosters, Despite Shortage', *PBS NewsHour*, 9 November 2009, http://www.pbs.org/newshour (accessed 20 December 2009).
12. In the 1976 vaccine campaign, 40 million Americans were immunized in 10 weeks. The campaign came to an abrupt halt in mid-December 1976, as incidences of Guillain-Barré syndrome began to occur in vaccinees. The Advisory Committee on Immunization Practices concluded in January 1977 that there was a slightly higher risk of Guillain-Barré in swine flu vaccinees (estimated at 1 per 100,000–200,000 vaccinations), which effectively

shut down the campaign. Richard E. Neustadt and Harvey V. Fineberg, *The Epidemic That Never Was: Policy-Making and the Swine Flu Scare* (New York: Random House, 1982), pp. 191–3. Cf. Richard E. Neustadt and Ernest R. May, *Thinking in Time: The Uses of History for Decision Makers* (New York: The Free Press, 1986).

13. Susan Peterson, 'Human Security, National Security, and Epidemic Disease', Ed. Robert L. Ostergard, Jr., *HIV/AIDS and the Threat to National and International Security* (Houndmills, Basingstoke: Palgrave Macmillan, 2007), p. 44. Catherine Boone and Jake Batsell, 'Politics and AIDS in Africa: Research Agendas in Political Science and International Relations', Robert L. Ostergard, Jr. (ed.), *HIV/AIDS and the Threat to National and International Security* (Houndmills: Palgrave Macmillan, 2007), p. 12.

14. Jonathan B. Tucker, 'Updating the International Health Regulations', *Biosecurity and Bioterrorism: Biodefense Strategy, Practice, and Science*, 3.4 (2005), pp. 346 (338–47). Mark W. Zacher and Tania J. Keefe, *The Politics of Global Health Governance: United by Contagion* (Houndmills, Basingstoke: Palgrave Macmillan, 2008), pp. 64–5.

# Bibliography

## Unpublished primary sources

*League of Nations Archives, Geneva.*
*League of Nations Health Committee.*
'Minutes of the Tenth Session of the Health Committee, April 26–27, 1927'.
*League of Nations Health Organisation.*
'Note on the Quinine Position in India', 1932. 8C Health, R5961, 27535/21313.
C.H. 687, 'Resolutions Adopted by the Advisory Council of the Eastern Bureau at its Session Held in New Delhi from the 26th to the 29th Dec, 1927'.
C.H./Malaria/135 in 8C Health, R5949, 3858/477.
C.H./Malaria/257, 21 Mar. 1938.
*League of Nations Secretariat, Health and Social Questions Section (1919–1932).*
12B Health, R991, 62193/57535.
12B Health, R991, 62193/58870.
8A Health, R6080, 40421/2652.
8B Health, R6157, 3031/378.
8C Health, R5953, 25128/1411.
8C Health, R6164, 903/903.
8C Health, R6173, 34241, 1561.
Letter from Biraud to Mhatre, 19 Nov. 1935. 8A Health, R6080, 2964/2652.
Letter from Ciuca to Christophers, 20 Oct. 1930. 8C Health, R5953, 23526/1911.
Letter from Graham to Rajchman, 11 Mar. 1933. 8C Health, R6163, 1242/1136.
Letter from Rajchman to Graham, 20 Jan. 1933. 8C Health, R6165, 911/911.
Letter from Rajchman to Russell, 28 Dec. 1934. 8C Health, R6166, 14000/911.
Letter from Russell to Rajchman, 17 Dec. 1934. 8A Health, R6062, 9361/985.
Letter from Russell to Rajchman, 13 Dec. 1938. 8A Health, R6080, 40421/2652.
Letter from Wilson to Ciuca, 15 Aug. 1939. 8C Health, R6174, 34241/1561.
*World Health Organization Archives, Geneva.*
ARC001, Records of the Office International d'Hygiène Publique, 1907–1946, Microfilm: T12.
ARC007, Collection on Parasitology of the Documentation Centre Communicable Diseases, WHO Headquarters: India Malaria, *Records of the Malaria Survey of India* (1932?).

## Published official primary sources

'Annual Report of the Director of Public Health, Madras. 1931'.
'Annual Report of the Public Health Commissioner with the Government of India'. (1920–1940.)
'Annual Report of the Sanitary Commissioner with the Government of India'. (1891–1919.)

*East India: Accounts and Estimates. 1901–1902.* London: Darling and Son, 1901.
*East India: Accounts and Estimates. 1912–1913.* London: Darling and Son, 1912.
*East India: Accounts and Estimates. 1921–1922.* London: Eason and Son, 1921.
*East India: Accounts and Estimates. 1932–1933.* London: His Majesty's Stationery Office, 1932.
'H1N1 Flu: Monitoring the Nation's Response'. United States Senate Committee on Homeland Security and Governmental Affairs. 21 October 2009.
'HHS 2009 H1N1 Vaccine Development Activities'. https://www. medicalcountermeasures.gov. Accessed 13 December 2009.
'Papers Relating to the Outbreak of Bubonic Plague in India; with Statement showing the Quarantine and other restrictions recently placed upon Indian trade up to March 1897. Presented to Parliament'. London: Eyre and Spottiswoode, 1897.
Plague Deposit Collections, vol. 2, 1897–1899. Proceedings of the Government of India, Home (Sanitary).
Plague Deposit Collections, vol. 3, 1900–1902. Proceedings of the Government of India, Home (Sanitary).
Proceedings of the Government of Bengal, Local (Public Health), 1935–1936.
Proceedings of the Government of Bengal, Municipal (Medical), 1895–1899, 1901, 1907–1908, 1911.
Proceedings of the Government of Bombay, General (General), 1897–1898, 1909, 1911, 1919–1920, 1931.
Proceedings of the Government of Bombay, General (Medical), 1894, 1905, 1911.
Proceedings of the Government of India, Education (Health), 1930–1931.
Proceedings of the Government of India, Education (Sanitary), 1912–1914, 1916–21.
Proceedings of the Government of India, Education (Sanitary – Deposit), 1920.
Proceedings of the Government of India, Education and Health (Sanitary), A and B Proceedings, 1922.
Proceedings of the Government of India, Education, Health and Lands (Agriculture), A Proceedings, 1926.
Proceedings of the Government of India, Education, Health and Lands (Education), A Proceedings, 1932–1933.
Proceedings of the Government of India, Education, Health and Lands (Health), 1926, 1929, 1931, 1934, 1940.
Proceedings of the Government of India, Home (Sanitary), 1894–1906, 1908, 1910.
Proceedings of the Government of Madras, Local (Public Health), 1930, 1933, 1936.
Proceedings of the Government of Madras, Local and Municipal (Plague), 1898–1900.
Proceedings relating to Quarantine. Vol. VII. 1892–1896. Proceedings of the Government of India, Home (Sanitary).
'Report (1939) of the Sub-Committee appointed by the Central Advisory Board of Health to examine the possibility of introducing a system of compulsory inoculation of pilgrims against cholera'. Simla: Government of India Press, 1940.
'Report of the Bombay Plague Committee (for period July 1, 1897–April 30, 1898)'. Bombay: Times of India Steam Press, 1898.

'Report of the Committee on the Organisation of Medical Research under the Government of India'. Calcutta: Government of India Central Publication Branch, 1929.

'Report of the Indian Delegation to the Secretary of State for India'.

'Report of the National Planning Committee. 1938'. New Delhi: Indian Institute of Applied Political Research, 1988 [1938].

'Report of the Pilgrim Committee. Madras, 1915'. Simla: Government Monotype Press, 1916.

'Reports on Native Papers: Bombay Presidency', 4.23, 17.9 (1897).

'Resolution of the Government of India in the Home Department. Sanitary. No. 227–240. Dated the 3rd February 1898'. *The Madras Plague Regulations and Rules*. Madras: Government Press, 1898.

*Return of the Budget of the Governor General of India in Council for 1939–40*. London: His Majesty's Stationery Office, 1939.

*Statistical Abstract Relating to British India from 1881–82 to 1890–91*. London: Eyre and Spottiswoode, 1892.

'Summary Report of the Inaugural Meeting of the Central Advisory Board of Health. Held in Simla on 22nd and 23rd June 1937'. Simla: Government of India Press, 1937.

'Travel Alert: China'. US Department of State, Bureau of Consular Affairs. 25 September 2009. http://travel.state.gov. Accessed 4 November 2009.

'View of the Government of India on the Memoranda submitted by Sir Ronald Ross and Colonel W. G. King, I. M. S., regarding the Prevention of Malaria in India'. Proceedings of the Government of India, Education (Sanitary), Jan. 1912. A. 32–33.

## Published unofficial primary sources

Abt, G. *Vingt-Cinq Ans d'Activité de L'Office International D'Hygiène Publique. 1909–1933*. Paris: Office International D'Hygiène Publique, 1933.

*Annals of the Indian Tea Association for 1915*. Calcutta: Indian Tea Association, 1916.

[Anonymous] 'A Visit to the Cinchona Plantations, Bengal'. *Indian Medical Gazette*, 47 (1912): 289–91.

[Anonymous] 'Abstract of the Annual Report for 1935 of the Public Health Commissioner with the Government of India'. *Indian Medical Gazette*, 73 (1938): 248.

[Anonymous] 'Annual Report of the All-India Institute of Hygiene and Public Health for the Year 1934'. *Indian Medical Gazette*, 70 (1935): 535–6.

[Anonymous] 'Annual Report of the Public Health Commissioner for the Year 1937'. *Indian Medical Gazette*, 73 (1938): 251.

[Anonymous] 'Annual Report of the Public Health Commissioner with the Government of India for 1936'. *Indian Medical Gazette*, 73 (1938): 708.

[Anonymous] 'Annual Report of the Public Health Commissioner with the Government of India for 1937. Vol. 1'. *Indian Medical Gazette*, 74 (1939): 774–5.

[Anonymous] 'Cholera and Bengal'. *Indian Medical Gazette*, 68 (1933): 521–2.

[Anonymous] 'Cinchona Policy'. *Indian Medical Gazette*, 67 (1932): 392–3.

[Anonymous] 'Foreign Opinion on Cholera'. *Indian Medical Gazette*, 27 (1892): 288.

[Anonymous] 'India and the League of Nations'. *Indian Medical Gazette*, 63 (1928): 159–60.

[Anonymous] *Indian Medical Gazette*, 73 (1938): 354–5.

[Anonymous] 'Indian Medical Research'. *Indian Medical Gazette*, 73 (1938): 230.

[Anonymous] 'Is Public Health Worth While'? *Indian Medical Gazette*, 65 (1930): 708.

[Anonymous] 'League of Nations Health Delegation'. *Modern Review*, 43.2 (1928): 235.

[Anonymous] 'Lord Sandhurst's Measures Against Plague'. *Indian Medical Gazette*, 32 (1897): 102.

[Anonymous] 'Major Bannerman. I. M. S. on the Results of Four Years' Inoculations Against Plague'. *Indian Medical Gazette*, 36 (1901).

[Anonymous] 'Malnutrition and Malaria'. *Modern Review*, 30.2 (1921): 266.

[Anonymous] 'Medical and Sanitary Matters in India'. *Indian Medical Gazette*, 30 (1895): 349.

[Anonymous] 'Medical Research in India'. *Indian Medical Gazette*, 52 (1917): 165.

[Anonymous] 'Note on the Quinine Policy of India by the Malaria Survey of India'. *Twelfth Conference of Medical Research Workers held at Calcutta from 26th November to 1st December 1934*. Simla: Government of India Press, 1935.

[Anonymous] 'Notes Epidemiologiques sur la Peste dans l'Inde', *Bulletin de l'Office International d'Hygiène Publique*. (May 1924): 585.

[Anonymous] 'Notification of Dangerous Diseases'. *Indian Medical Gazette*, 64 (1929).

[Anonymous] 'Plague Prevention in Bengal'. *Indian Lancet*, (16 September 1897): 287–8.

[Anonymous] 'Presidential Address in Public Health'. *Indian Medical Gazette*, 30 (1895): 18.

[Anonymous] 'Preventive Medicine a Factor in Empire Building'. *Indian Medical Gazette*, 39 (1904): 382.

[Anonymous] 'Proceedings of the Imperial Malaria Committee held in Bombay on 16th and 17th November 1911'. *Paludism, Being the Transactions of the Committee for the Study of Malaria in India*, 4 (1912): 2.

[Anonymous] 'Rats and Plague'. *Modern Review*, 43.6 (1928): 765–6.

[Anonymous] 'Report of the Health Organisation'. *Quarterly Bulletin of the Health Organisation*, (September 1932): 117.

[Anonymous] 'Report on the Session of the Office International D'Hygiène Publique, held in Paris from 25th April to 4th May. 1932'. *Indian Medical Gazette*, 68 (1933): 119–20.

[Anonymous] 'Research in Tropical Diseases in India'. *Indian Medical Gazette*, 40 (1905): 307.

[Anonymous] 'Sanitation in the Two Bengals'. *Indian Medical Gazette*, 46 (1911): 361.

[Anonymous] 'The Cape Town Conference'. *Quarterly Bulletin of the Health Organisation of the League of Nations*, 2 (1933): 22.

[Anonymous] 'The Cholera Danger in India: Melas and the Spread of Cholera'. *Indian Medical Gazette*, 74 (1939): 489.

[Anonymous] 'The Economic Factor in Tropical Diseases'. *Indian Medical Gazette*, 57 (1922): 342–3.

[Anonymous] 'The Far Eastern Association of Tropical Medicine'. *Indian Medical Gazette*, 62 (1927): 280.

[Anonymous] 'The Future of Malaria Control in India'. *Indian Medical Gazette*, 62 (1927): 28–33.

[Anonymous] 'The Malaria Policy'. *Indian Medical Gazette*, 66 (1931): 520.

[Anonymous] 'The Menace of Malaria'. *Modern Review*, 40.4 (1926): 463.

[Anonymous] 'The Need for a Public Health for India'. *Indian Medical Gazette*, 62 (1927): 575.

[Anonymous] 'The New Antiplague Campaign'. *Indian Medical Gazette*, 42 (1907): 381–2.

[Anonymous] 'The Office International d'Hygiène Publique'. *Indian Medical Gazette*, 65 (1930): 582.

[Anonymous] 'The Paternal *versus* the Common Sense Plague Policy'. *Indian Medical Gazette*, 39 (1904): 447–50.

[Anonymous] 'The Plague Commission Report'. *Indian Medical Gazette*, 35 (1900): 141–2.

[Anonymous] 'The Rockefeller Foundation, New York: Annual Report, 1936'. *Indian Medical Gazette*, 73 (1938): 184.

[Anonymous] 'The Suppression of Plague and Malaria in India'. *Indian Medical Gazette*, 46 (1911): 429–30.

[Anonymous] 'The Therapeutics of Malaria'. *Quarterly Bulletin of the Health Organisation of the League of Nations*, 2 (1933): 275.

[Anonymous] 'The Yellow Fever Peril'. *Indian Medical Gazette*, 74 (1939): 698–9.

[Anonymous] 'Travaux de la Commission Anglaise de la Peste aux Indes'. *Bulletin de l'Office International d'Hygiène Publique* (January 1909.)

[Anonymous] 'Yellow Fever: Transmission over Continents'. *Times*, 24 February 1933.

Bahadur, Rai. and G. C. Chatterji. 'Biological Controls of Malaria in the Rural Areas of Lower Bengal'. *Modern Review*, 54.6 (1933): 647–57.

Barraud, P. J. 'The Distribution of "Stegomyia Fasciata" in India with Remarks on Dengue and Yellow Fever'. *Indian Journal of Medical Research*, 16 (1928–29): 377–86.

Bentley, C. A. 'Propagande en faveur de la quinisation au Bengale'. *Bulletin de l'Office International d'Hygiène Publique* (October 1913): 1864.

——. *Report of an Investigation into the Causes of Malaria in Bombay and the Measures Necessary for its Control*. Bombay: Government Central Press, 1911.

——. 'Quinine Propaganda'. *Proceedings of the Third Meeting of the General Malaria Committee held at Madras November 18, 19 and 20, 1912*. Simla: Government Central Branch Press, 1913.

——. 'Une nouvelle conception du paludisme'. *Bulletin de l'Office International d'Hygiène Publique* (October 1910): 1863.

Bradfield, E. W. C. 'Notes on Medical and Public Health Organisation in the Bombay Presidency'. *Intergovernmental Conference of Far-Eastern Countries on Rural Hygiene. Preparatory Papers relating to British India*. Geneva: League of Nations, 1937.

——. *An Indian Medical Review*. New Delhi: Government of India Press. 1938.

British Social Hygiene Council. *Empire Social Hygiene Year-Book. 1937*. London: George Allen and Unwin, 1937.

Buchanan, George, S. 'Letters to the Editor: Yellow Fever. Transmission by Aircraft'. *Times*, 25 February 1933.

Calder, C. C. 'Letter from C. C. Calder, Director of the Botanical Survey to R. Littlehailes, Educational Commissioner with the Government of India, 4 Jan 1932'. *Proceedings of the Government of India, Education, Health and Lands* (Education), 14–8/32, 1932.

Chatterjee, Sailaj Lal. 'Anti-Mosquito Measures'. *Calcutta Municipal Gazette*, 3.19 (1926): 815.

Chatterji, A. C. 'Note on Public Health Organisation in Bengal'. *Intergovernmental Conference of Far-Eastern Countries on Rural Hygiene. Preparatory Papers relating to British India*. Geneva: League of Nations, 1937.

Christophers, S. R. and W. F. Harvey. 'Malaria Research and Preventive Measures Against Malaria in the Federated Malay States and in the Dutch East Indies'. *Indian Journal of Medical Research*, 10 (1922–23): 771.

——. *Malaria in the Duars*. Simla: Government Monotype Press, 1911.

Clemesha, Wm. Wesley. *Plague. From the Sanitarian's Point of View*. Calcutta: Baptist Mission Press, 1903.

Clemow, Frank G. *The Geography of Disease*. Cambridge: University Press, 1903.

Condon, J. K. *The Bombay Plague: Being a History of the Progress of Plague in the Bombay Presidency from September 1896 to June 1899*. Bombay: Education Society, 1900.

*Conference Held at Venice in January 1892 Respecting the Sanitary Regulations of Egypt. Presented to both Houses of Parliament by Command of Her Majesty. June 1892*. London: Harrison and Sons, 1892.

*Conférence Sanitaire Internationale de Paris. 10 Octobre–3 Decembre 1903. Procès-Verbaux.* Paris: Imprimerie Nationale, 1904.

*Conférence Sanitaire Internationale de Paris. 7 Novembre 1911–17 Janvier 1912. Procès-Verbaux.* Paris: Imprimerie Nationale, 1912.

*Conférence Sanitaire Internationale de Paris. 7 Février–3 Avril. 1894. Procès-Verbaux.* Paris: Imprimerie Nationale, 1894.

*Conférence Sanitaire Internationale de Venise. 16 Février–19 Mars 1897. Procès-Verbaux.* Rome: Forzani, 1897.

*Conférence Sanitaire Interrnationale de Paris (10 Mai–21 Juin 1926. Procès-Verbaux.* Paris: Imprimerie Nationale, 1927.

Cook, Nield. 'Plague Precautions'. *Indian Lancet*, (16 May 1898): 479–82.

Covell, G. 'A Note on the Method Used to Combat Rural Malaria in India'. *Intergovernmental Conference of Far-Eastern Countries on Rural Hygiene: Preparatory Papers relating to British India*. Geneva: League of Nations, 1937.

——. 'Foreward' in R. K. Mhatre. *A Survey of Aedes Mosquitoes in Bombay and the Measures Suggested for Their Control*. Bombay: British India Press, 1934.

——. 'The Malaria Survey of India. 1927–1937', *The Journal of the Malaria Institute of India* (March 1938): 2–11.

——. *Malaria in Bombay*. Bombay: Government Central Press, 1928.

Coyajee, J. C. *India and the League of Nations*. Madras: Waltair, 1932.

Crombie, A. 'Haffkine's Anti-Choleraic Inoculation'. *Medical Reporter* (16 September 1895): 173.

Fry, A. B. 'Le Choléra dans le Bengale. dans le Passé et à l'Heure Actuelle'. *Bulletin de l'Office International d'Hygiène Publique*, 18.3 (1926): 299.

Ganapathy, C. M. 'Public Health Administration in Madras Presidency'. *Intergovernmental Conference of Far-Eastern Countries on Rural Hygiene: Preparatory Papers relating to British India*. Geneva: League of Nations. 1937.

Gangulee, N. *Health and Nutrition in India*. London: Faber and Faber, 1939.

Ghosh, Dakshina, R. 'How to Fight Malaria in Our Villages'. *Modern Review*, 19.2 (1916): 199–201.

Gill, C. A. 'The Theory and Practice of Malaria "Control."' *Far Eastern Association of Tropical Medicine. Transactions of the Seventh Congress. British India 1927.* Vol. 2. Calcutta: Thacker's Press. 1928.

Graham, J. D. 'Hygiene and Public Health'. *Indian Medical Gazette: Supplement – The Indian Medical Year. 1926. A Review* (1927): 16.

———. 'International Aspects of Disease with Special Reference to Quarantine', *Far Eastern Association of Tropical Medicine. Transactions of the Seventh Congress. British India 1927.* Vol. 1. Calcutta: Thacker's Press, 1928.

———. 'La Peste dans l'Inde Britannique'. *Bulletin de l'Office International d'Hygiène Publique*, 22.11 (1930): 2088–9.

Haffkine, W. M. and Surgeon-Major Lyons. 'Joint Report on the Epidemic of Plague in Lower Damaun. Portuguese India. and on the Effect of Preventive Inoculation There'. *Indian Lancet*, 8.12 (1897): 594–5.

Haffkine, W. M. 'A Lecture on Vaccination Against Cholera'. *Indian Lancet* (16 February 1896).

———. 'On the Present Methods of Combating the Plague'. *Proceedings of the Royal Society of Medicine* (Epidemiological Section) 1.1 (1908): 79–80.

*Health Organisation in British India*. Calcutta: Thacker's Press, 1928.

Hehir, Patrick. *Malaria in India*. London: Oxford University Press, 1927.

Holt, Joseph. *An Epitomized Review of the Principles and Practice of Maritime Sanitation*. New Orleans: L. Graham & Son, 1892.

Hoops, A. L. *Present Day Public Health in India: A Report on the League of Nations Interchange of Health Officers in India (1st January-18th February, 1928)*. London: John Bale, Sons, and Danielsson, 1928.

*Indian Social Reformer* (3 August 1929) in LON. 8C Health. R5949. 503/503.

James, S. P. 'Connaissances Récemment Acquises sur la Fièvre Jaune'. *Bulletin Mensuel de l'Office International D'Hygiène Publique*, 25 (1933): 62–63.

King, W. G. 'The Prevention of Malaria in India' in Proceedings of the Government of India, Education (Sanitary), January 1912.

———. 'The Protection of India from Yellow Fever'.

League of Nations Health Organization. *International Health Year-Book. 1925.* Vol. 2. Geneva: League of Nations, 1926.

Leslie, J. T. W. 'An Address on Malaria in India'. *Lancet*, 174 (1909): 1483–4.

———. 'Malaria in India'. *Proceedings of the Imperial Malaria Conference held at Simla in October 1909*. Simla: Government Central Branch Press. 1910.

Liston, William Glen. 'The Cause and Prevention of the Spread of Plague in India. A lecture delivered before the Bombay Sanitary Association on 11th December 1907'.

———. 'Plague Preventive Measures' in *The Proceedings of the Second All-India Sanitary Conference held at Madras in November 1912*. Vol. 3. Simla: Government Central Branch Press, 1913.

Megaw, J. W. D. *Confidential – Further Note on the Formation of a Public Health Board*. Simla: Government of India Press, 1932.

Mhatre, R. K. *A Survey of Aedes Mosquitoes in Bombay and the Measures Suggested for Their Control*. Bombay: British India Press, 1934.

*Ninth Conference of Medical Research Workers held at Calcutta from 30th November to 5th December 1931*. Simla: Government of India Press, 1932.

*Paludism, Being the Transactions of the Committee for the Study of Malaria in India.* 3 (1911): 10–14.

'Proceedings of the Government of India, Sanitary (Plague), Apr. 1899'. *Report of the Committee of the Bengal Chamber of Commerce: From 1st February 1899 to 31st January 1900.* Vol. 2. Calcutta: W. Newman & Company, 1900.

Ram, V. Shiva, and Brij Mohan Sharma. *India and the League of Nations.* Lucknow: Upper India Publishing House, 1932.

*Report of the Bombay Chamber of Commerce for the Year 1897.* Bombay: Bombay Gazette Steam Printing Works, 1898.

*Report of the Bombay Chamber of Commerce for the Year 1899.* Bombay: Bombay Gazette Steam Printing Works, 1900.

*Report of the Intergovernmental Conference of Far-Eastern Countries on Rural Hygiene. Held at Bandoeng (Java. August 3–13. 1937*, Geneva: League of Nations, 1937.

*Report of the Malaria Commission on its Study Tour in India.* Geneva: League of Nations, 1930.

Ross, Ronald. 'A Memorandum on the Present Position of Malaria-Prevention in India'. Proceedings of the Government of India, Education (Sanitary), January 1912.

——. 'Indian Fevers' in *Philosophies.* London: John Murray, 1911.

——. *Memoirs with a Full Account of the Great Malaria Problem and its Solution.* New York: E. P. Dutton, 1923.

Rüffer, Juan Armand. 'Letter from Juan Armand Rüffer, British consul in Alexandria, to Lord Cromer'. Proceedings of the Government of India, Home (Sanitary), September 1900.

Russell, A. J. H. 'A Note on the Central Government's Health Organization and Associated Institutions and Organisations Concerned with Public Health'. *Intergovernmental Conference of Far-Eastern Countries on Rural Hygiene. Preparatory Papers relating to British India.* Geneva: League of Nations, 1937.

——. 'Cholera in India'. *Far Eastern Association of Tropical Medicine. Transactions of the Ninth Congress held at Nanking. 1934.* Vol. 1. Nanking: National Health Administration, 1935?

——. 'Périodicité du Cholera dans l'Inde'. *Bulletin de l'Office International d'Hygiène Publique,* 17.8 (1925): 901.

——. 'Plague in India'. *Far Eastern Association of Tropical Medicine. Transactions of the Ninth Congress held at Nanking. 1934.* Vol. 2. Nanking: National Health Administration 1935?

——. 'Population and Public Health in India', *Far Eastern Association of Tropical Medicine. Transactions of the Seventh Congress. British India 1927.* Vol. 1. Calcutta: Thacker's Press, 1928.

Russell, Paul F. 'Malaria in India: Impressions from a Tour'. *The American Journal of Tropical Medicine,* 16 (1936): 655.

Schüffner, W. A. P. 'Le Paludisme aux Indes Britanniques'. *Compte-Rendu du Deuxième Congrès International du Paludisme et de la Célébration du Cinquantenaire del Découverte de Laveran.* Alger: Institut Pasteur, 1931.

Simpson, W. J. 'Maritime Quarantine and Sanitation in Relation to Cholera'. *The Pratitioner: A Journal of Therapeutics and Public Health,* 48 (1892): 148–60.

Sinton, J. A. 'Suggestions with regard to the prevention of the spread of Yellow Fever to India by Air Traffic with special reference to Insect Transmission'. *Health Bulletin,* 20 (1934): 9.

——. *What Malaria Costs India*, Delhi: Government of India Press, 1956 [1939].

Sinton, J. A. and Raja Ram. *Man-Made Malaria in India*. Simla: Government of India Press, 1938.

Snijders, E. P. 'The Yellow Fever Problem in the Far East'. *Far Eastern Association of Tropical Medicine: Transactions of the Eighth Congress held in Siam 1930*. Vol. 1. Bangkok: Bangkok Times Press, 1931.

Sprawson, C. A. 'The Position of India in Regard to the Yellow Fever Question'. *Quarterly Bulletin of the Health Organisation*, 5 (1936): 87–8.

Strickland, C. F. *Quinine and Malaria in India*. London: Oxford University Press, 1939.

Taylor, J. 'Rural Plague in India'. *Intergovernmental Conference of Far-Eastern Countries on Rural Hygiene: Preparatory Papers relating to British India*. Geneva: League of Nations, 1937.

*The Proceedings of the First All-India Sanitary Conference held at Bombay on 13th and 14th November 1911*. Calcutta: Superintendent Government Printing, 1912.

*The Proceedings of the Second All-India Sanitary Conference held at Madras in November 1912*. Vol. 1. Simla: Government Central Branch Press, 1913.

*The Proceedings of the Third All-India Sanitary Conference held at Lucknow in January 1914 and Resolution of the Government of India on Sanitation in India*. London: His Majesty's Stationary Office, 1914.

*The Treatment of Malaria: Study of Synthetic Drugs. As compared with Quinine in the Therapeutics and Prophylaxis of Malaria*. Geneva: League of Nations, 1937.

Turner, J. A. *Sanitation in India*. Bombay: Times of India. 1914.

*Twelfth Conference of Medical Research Workers held at Calcutta from 26th November to 1st December 1934*. Simla: Government of India Press, 1935.

Twiss, Travers. *The Law of Nations considered as Independent Political Communities*. Oxford: University Press, 1861 [reprinted by Gaunt. 2000].

Watson, Malcolm. 'Observations on Malaria Control with Special Reference to the Assam Tea Gardens and Some Remarks on Mian Mir Lahore Cantonment'. *Transactions of the Royal Society of Tropical Medicine and Hygiene*. 18.4 (1924) 152.

Wellington, A. R. *Hygiene and Public Health in India. Report on Conditions Met with During the Tour of the League of Nations Interchange of Health Officers*. Kuala Lumpur: Federated Malay States Government Press, 1929.

White, F. Norman. *The Prevalence of Epidemic Disease and Port Health Organisation and Procedure in the Far East*. Geneva: League of Nations, 1923.

——. *Twenty Years of Plague in India with Special Reference to the Outbreak of 1917–18*. Calcutta: Superintendent Government Printing, 1918.

Wilkinson, E. 'A Revised Scheme for the Distribution of Quinine by Government'. *Proceedings of the Imperial Malaria Conference held at Simla in October 1909*. Simla: Government Central Branch Press, 1910.

——. 'Report on Inquiries into the Measures for the Sanitary Control of the Hejaz Pilgrimage, 1919'.

## Published and unpublished secondary sources

'PEPFAR Funding'. http://www.pepfar.gov. Accessed 2 January 2012.

Adams, John. and Robert Craig West. 'Money, Prices, and Economic Development in India, 1861–1895', *Journal of Economic History*, 39.1 (1979): 55–68.

Agarwal, R. C. *Constitutional Development and National Movement of India.* New Delhi: S. Chand, 1991.

Aginam, Obijiofor. 'International Law and Communicable Diseases'. *Bulletin of the World Health Organization* (2002): 946–51.

Amrith, Sunil. 'Rockefeller Foundation and Postwar Public Health in India'. http:// archive.rockefeller.edu/publications/resrep/amrith.pdf. Accessed 14 May 2007.

——. *Decolonizing International Health: India and Southeast Asia, 1930–65.* Houndmills, Basingstoke: Palgrave Macmillan, 2006.

Anderson, Warwick. 'Postcolonial Histories of Medicine'. *Locating Medical History: The Stories and Their Meanings.* Frank Huisman and John Harley Warner (eds). Baltimore: The Johns Hopkins University Press, 2004.

Annan, Kofi. 'Secretary-General Proposes Global Fund for Fight Against HIV/ AIDS and Other Infectious Diseases at African Leaders Summit'. http://www. theglobalfund.org/en/history. Accessed 8 December 2008.

Armitage, David. 'Three Concepts of Atlantic History'. *The British Atlantic World, 1500–1800.* David Armitage and Michael J. Braddick (eds). Houndmills, Basingstoke: Palgrave Macmillan, 2002.

Arnold, David. '"An Ancient Race Outworn": Malaria and Race in Colonial India, 1860–1930', *Race, Science and Medicine, 1700–1960.* Bernard Harris and Waltraud Ernst (eds). London: Routledge, 1999, 128.

——. 'Cholera and Colonialism in British India'. *Past and Present*, 113 (1986): 118–51.

——. 'Introduction: Tropical Medicine before Manson'. *Warm Climates and Western Medicine: The Emergence of Tropical Medicine, 1500–1900.* David Arnold (ed.). Amsterdam: Rodopi B.V., 1996.

——. 'Public Health and Public Power: Medicine and Hegemony in Colonial India'. *Contesting Colonial Hegemony: State and Society in Africa and India.* Dagmar Engels and Shula Marks (eds). London: British Academic Press, 1994.

——. 'The "Discovery" of Malnutrition and Diet in Colonial India'. *The Indian Economic and Social History Review*, 31.1 (1994): 1–26.

——. 'The Indian Ocean as a Disease Zone, 1500–1950'. *South Asia* (1991): 1–21.

——. 'Touching the Body: Perspectives on the Indian Plague, 1896–1900'. *Selected Subaltern Studies.* Ranajit Guha and Gayatri Chakravorty Spivak (eds). Oxford: Oxford University Press, 1988.

——. *Colonizing the Body: State Medicine and Epidemic Disease in Nineteenth-Century India.* Berkeley: University of California Press, 1993.

——. *Science, Technology and Medicine in Colonial India.* Cambridge: Cambridge University Press, 2000.

Audi, Nadim. 'Culling Pigs in Flu Fight, Egypt Angers Herders and Dismays UN'. *New York Times*, 30 April 2009. http://www.nytimes.com. Accessed 4 November 2009.

Bailyn, Bernard. *Atlantic History: Concept and Contours.* Cambridge: Harvard University Press, 2005.

Bala, Poonam (ed.) *Biomedicine as a Contested Site: Some Revelations in Imperial Contexts.* Lanham: Lexington Books, 2009.

Balachandran, G. 'Reappraisal: Finance and Politics in Late Colonial India 1917–1947'. *South Asia*, 19.1 (1996): 77–101.

——. 'Introduction'. *India and the World Economy, 1850–1950.* G. Balanchandran (ed.). New Delhi: Oxford University Press, 2003.

Baldwin, Peter. *Contagion and the State in Europe, 1830–1930*. Cambridge: Cambridge University Press, 1999.

Ballantyne, Tony. 'Empire, Knowledge and Culture: From Proto-Globalization to Modern Globalization'. *Globalization in World History*. A. G. Hopkins (ed.). New York: W.W. Norton & Company, 2002.

——. *Orientalism and Race: Aryanism in the British Empire*. London: Palgrave Macmillan, 2002.

Bardi, Jason Socrates. 'Grounding Flights Won't Stop Flu'. *Live Science*, 1 October 2009. http://www.livescience.com. Accessed 4 November 2009.

Bashford, Alison. '"The Age of Universal Contagion": History, Disease and Globalization'. *Medicine at the Border: Disease Globalization and Security, 1850 to the Present*. Alison Bashford (ed.). New York: Palgrave Macmillan, 2006.

——. 'Global Biopolitics and the History of World Health'. *History of the Human Sciences*, 19.1 (2006): 67–88.

——. *Imperial Hygiene: A Critical History of Colonialism. Nationalism and Public Health*. New York: Palgrave Macmillan, 2004.

Bashford, Alison (ed). *Medicine at the Border: Disease, Globalization and Security, 1850 to the Present*. New York: Palgrave Macmillan, 2006.

Bayly, C. A. '"Archaic" and "Modern" Globalization in the Eurasian and African Arena. c, 1750–1850'. *Globalization in World History*. A. G. Hopkins (ed.). New York: W.W. Norton, 2002.

——. *The Birth of the Modern World, 1780–1914: Global Connections and Comparisons*. Malden: Blackwell Publishing, 2004.

Bayly, C. A. and Leila Fawaz. 'Introduction'. *Modernity and Culture: From the Mediterranean to the Indian Ocean*. Leila Tarazi Fawaz, C. A. Bayly, and Robert Ilbert (eds). Columbia University Press, 2002.

Beck, Ulrich. *World Risk Society*. Cambridge: Polity Press, 1999.

Bell, Heather. *Frontiers of Medicine in the Anglo-Egyptian Sudan, 1899–1940*. Oxford: Clarendon Press, 1999.

Bennett, Peter. 'Understanding Responses to Risk: Some Basic Findings'. *Risk Communication and Public Health*. Peter Bennett and Sir Kenneth Calman (eds). Oxford: Oxford University Press, 1999.

Boone, Catherine and Jake Batsell. 'Politics and AIDS in Africa: Research Agendas in Political Science and International Relations'. *HIV/AIDS and the Threat to National and International Security*. Robert L. Ostergard, Jr. (ed.). Houndmills, Basingstoke: Palgrave Macmillan, 2007.

Bose, Sugata. *Agrarian Bengal: Economy, Social Structure and Politics, 1919–1947*. Cambridge: Cambridge University Press, 1986.

——. 'Instruments and Idioms of Colonial and National Development: India's Historical Experience in Comparative Perspective', *International Development and the Social Sciences: Essays on the History and Politics of Knowledge*. Frederick Cooper and Randall Packard (eds). Berkeley: University of California Press, 1997.

——. 'Space and Time on the Indian Ocean Rim: Theory and History'. *Modernity and Culture: From the Mediterranean to the Indian Ocean*. Leila Tarazi Fawaz, C. A. Bayly, and Robert Ilbert (eds). Columbia University Press, 2002.

——. *A Hundred Horizons: the Indian Ocean in the Age of Global Empire*. Cambridge: Harvard University Press, 2006.

——. *Peasant Labour and Colonial Capital: Rural Bengal Since 1770*. Cambridge: Cambridge University Press, 1993.

Bose, Sugata. and Ayesha Jalal. *Modern South Asia: History, Culture, Political Economy*. New York: Routledge, 1998 (First edition), 2004 (Second edition).

Brandt, Allan, M. 'Behavior, Disease, and Health in the Twentieth-Century United States: The Moral Valence of Individual Risk'. *Morality and Health*. Allan M. Brandt and Paul Rozin (eds). (New York and London: Routledge, 1997), p. 54.

——. 'From Analysis to Advocacy: Crossing Boundaries as a Historian of Health Policy'. *Locating Medical History: The Stories and Their Meanings*. Frank Huisman and John Harley Warner (eds). Baltimore: The Johns Hopkins University Press, 2004.

——. '"Just Say No": Risk, Behavior, and Disease in Twentieth-Century America'. *Scientific Authority and Twentieth-Century America*. Ronald G. Walters (ed.). Baltimore: The Johns Hopkins University Press, 1997.

——. *The Cigarette Century: The Rise Fall, and Deadly Persistence of the Product that Defined America*. New York: Basic Books, 2007.

Brown, Judith, M. 'Imperial Facade: Some Constraints upon and Contradictions in the British Position in India, 1919–35'. *Transactions of the Royal Historical Society* Series 5. 26 (1976): 35–52.

Buckingham, Jane. *Leprosy in Colonial South India: Medicine and Confinement*. New York: Palgrave Macmillan, 2002.

Bull, Hedley. *The Anarchical Society: A Study of Order in World Politics*. New York: Columbia University Press, 1977.

Bynum, W. F. 'Policing Hearts of Darkness: Aspects of the International Sanitary Conferences'. *History and Philosophy of the Life Sciences*, 15 (1993): 421–34.

——. *Science and the Practice of Medicine in the Nineteenth Century*. Cambridge: Cambridge University Press, 1994.

Cain, P. J. and A. G. Hopkins. *British Imperialism, 1688–2000*. Harlow: Longman, 2002.

Cannadine, David. 'The Empire Strikes Back'. *Past and Present* (1995): 180–94.

Carr, Edward Hallett. *The Twenty Years' Crisis, 1919–1939*. London: Macmillan & Co., 1954.

Carty, Daniel. 'Obama: Swine Flu 'Not A Cause For Alarm'. *CBS News*, 27 April 2009. http://www.cbsnews.com. Accessed 6 November 2009.

Catanach, I. J. '"Fatalism"? Indian Responses to Plague and Other Crises'. *Asian Profile*, 12.2 (1984): 183–92.

——. 'Plague and the Tensions of Empire: India 1896–1918'. *Imperial Medicine and Indigenous Societies*. David Arnold (ed.). Manchester: Manchester University Press, 1988.

Chakrabarty, Dipesh. *Habitations of Modernity: Essays in the Wake of Subaltern Studies*. Chicago: The University of Chicago Press, 2002.

——. *Provincializing Europe: Postcolonial Thought and Historical Difference*. Princeton: Princeton University Press, 2000.

Chamberlain, Muriel, E. *The Formation of the European Empires, 1488–1920*. Harlow: Pearson Education, 2000.

Chandavarkar, Rajnarayan. 'Plague Panic and Epidemic Politics in India, 1896–1914'. *Epidemics and Ideas: Essays on the Historical Perception of Pestilence*. Terence Ranger and Paul Slack (eds). Cambridge: Cambridge University Press, 1992.

Chatterjee, Partha. *The Nation and Its Fragments: Colonial and Postcolonial Histories*. Princeton: Princeton University Press, 1993.

——. *Nationalist Thought and the Colonial World: A Derivative Discourse.* Minneapolis: University of Minnesota Press, 1986.

Chaudhuri, K. N. *Trade and Civilisation in the Indian Ocean: An Economic History from the Rise of Islam to 1750.* Cambridge: Cambridge University Press, 1985.

Clarke, Lee and James F. Short, Jr. 'Social Organization and Risk: Some Current Controversies'. *Annual Review of Sociology*, 19 (1993): 375–99.

Cohen, Jon. 'Out of Mexico? Scientists Ponder Swine Flu's Origins'. *Science*, 324.5928: 700–2. www.sciencemag.org. Accessed 5 November 2009.

Cohn, Bernard. *Colonialism and its Forms of Knowledge: The British in India.* Princeton: Princeton University Press, 1996.

Cohrssen, John, J. and Vincent, T. Covello. *Risk Analysis: A Guide to Principles and Methods for Analyzing Health and Environmental Risks.* Springfield, VA: United States Council on Environmental Quality, 1989.

Coleman, William. *Death is a Social Disease: Public Health and Political Economy in Early Industrial France.* Madison: The University of Wisconsin Press, 1982.

Cooper, Frederick. *Colonialism in Question: Theory. Knowledge. History.* Berkeley: University of California Press, 2005.

Covello, Vincent, T. and Branden B. Johnson. 'The Social and Cultural Construction of Risk: Issues, Methods, and Case Studies'. *The Social and Cultural Construction of Risk: Essays on Risk Selection and Perception.* Branden B. Johnson and Vincent T. Covello (eds). Dordrecht: D. Reidel Publishing Company, 1987.

'Cruise Ship Pacific Dawn Quarantined near Cairns over Swine Flu'. *The Australian*, 27 May 2009. http://www.theaustralian.com.au. Accessed 5 November 2009.

Cunningham, Andrew. 'Transforming Plague: The Laboratory and the Identity of Infectious Disease'. *The Laboratory Revolution in Medicine.* Andrew Cunningham and Perry Williams (eds). Cambridge: Cambridge University Press, 1992.

Curtin, Philip. *The World and the West: The European Challenge and the Overseas Response in the Age of Empire.* Cambridge: Cambridge University Press, 2000.

Douglas, Mary and Aaron Wildavsky. *Risk and Culture: An Essay on the Selection of Technical and Environmental Dangers.* Berkeley: University of California Press, 1982.

Drayton, Richard. 'Science. Medicine and the British Empire'. *The Oxford History of the British Empire: Volume V: Historiography.* Robin W. Winks (ed.). Oxford: Oxford University Press, 1999.

——. *Nature's Government: Science. Imperial Britain and 'Improvement' of the World.* New Haven: Yale University Press, 2000.

Dumett, Raymond, E. 'The Campaign against Malaria and the Expansion of Scientific Medical and Sanitary Services in British West Africa, 1898–1910'. *African Historical Studies*, 1.2 (1968): 153–97.

Dutta, Achintya Kumar. '*Kala-Azar* in Assam: British Medical Intervention and People's Response'. *Maladies. Preventives and Curatives: Debates in Public Health in India.* Amiya Kumar Bagchi and Krishna Soman (eds). Kolkata: Tulika Books, 2005.

Dutta, Sanchari. 'Plague, Quarantine and Empire: British-Indian sanitary strategies in Central Asia, 1897–1907'. *The Social History of Health and Medicine in Colonial India.* Biswamoy Pati and Mark Harrison (eds). London: Routledge, 2009.

Echenberg, Myron. *Plague Ports: The Global Urban Impact of Bubonic Plague, 1894–1901*. New York: New York University Press, 2007.

El Deeb, Sarah. 'Egypt Warns of Post-Hajj Swine Flu Quarantine'. *abc News*, 20 May 2009. http://abcnews.go.com. Accessed 5 November 2009.

Espinosa, Mariola. *Epidemic Invasions: Yellow Fever and the Limits of Cuban Independence, 1878–1930*. Chicago: University of Chicago Press, 2009.

Evans, Hughes. 'European Malaria Policy in the 1920s and 1930s: The Epidemiology of Minutiae'. *Isis*, 80 (1989): 40–59.

Evans, Richard J. 'Epidemics and Revolutions: Cholera in Nineteenth-Century Europe'. *Past and Present*, 120.1 (1988): 123–46.

Eyler, John. *Sir Arthur Newsholme and State Medicine, 1885–1935*. Cambridge: Cambridge University Press, 1997.

——. *Victorian Social Medicine: the ideas and methods of William Farr*. Baltimore: The Johns Hopkins University Press, 1979.

Farley, John. 'Parasites and the Germ Theory of Disease'. *Framing Diseases: Studies in Cultural History*. Charles E. Rosenberg and Janet Golden (eds). New Brunswick: Rutgers University Press, 1992.

Fearnley, Lyle. 'Redesigning Syndromic Surveillance for Biosecurity'. *Biosecurity Interventions: Global Health and Security in Question*. Andrew Lakoff and Stephen J. Collier (eds). New York: Columbia University Press, 2008.

Fidler, David, P. 'The Globalization of Public Health: The First 100 Years of International Health Diplomacy'. *World Health Organization*, 79 (2001): 842–9.

——. *International Law and Infectious Diseases*. Oxford: Clarendon Press, 1999.

——. 'Public Health and International Law: The Impact of Infectious Diseases on the Formation of International Legal Regimes, 1800–2000'. *Plagues and Politics: Infectious Disease and International Policy*. Andrew T. Price-Smith (ed.). Houndmills, Basingstoke: Palgrave Macmillan, 2001.

——. 'Towards a Global *ius pestilentiae*: The Functions of Law in Global Biosecurity'. *Global Biosecurity: Threats and Responses*. Peter Katona, John P. Sullivan and Michael D. Intriligator (eds). London: Routledge, 2010.

Foucault, Michel. *The Birth of Biopolitics: Lectures at the Collège de France, 1978–1979*. Michel Senellart (ed.). New York: Picador, 2004.

——. *Security, Territory, Population: Lectures at the Collège de France, 1977–1978*. Michel Senellart (ed.). New York: Picador, 2004.

Frenk, Julio. 'Economic Crises and Health: Risk or Opportunity'? Lecture given at Center for History and Economics, Harvard University. Cambridge, 2 December 2008.

Giddens, Anthony. *Modernity and Self-Identity: Self and Society in the Late Modern Age*. Stanford: Stanford University Press, 1991.

Goodman, Neville M. *International Health Organizations and Their Work*. Philadelphia: The Blakiston Company, 1952.

Grove, Richard, H. *Green Imperialism: Colonial. Expansion. Tropical Island Edens and the Origins of Environmentalism, 1600–1860*. Cambridge: Cambridge University Press, 1995.

Guha, Ranajit. 'Dominance without Hegemony and Its Historiography'. *Subaltern Studies VI*. Ranajit Guha (ed.). Delhi: Oxford University, 1989.

Guha, Sumit. *Health and Population in South Asia: From Earliest Times to the Present*. London: Hurst, 2001.

Hamlin, Christopher. 'State Medicine in Great Britain'. *The History of Public Health and the Modern State*. Dorothy Porter (ed.). Amsterdam: Rodopi B.V., 1994.

——. *Public Health and Social Justice in the Age of Chadwick: Britain, 1800–1854*. Cambridge: Cambridge University Press, 1998.

Harremoës, Poul et al. 'Introduction'. *The Precautionary Principle in the 20th Century: Late Lessons from Early Warnings*. Poul Harremoës, David Gee, Malcolm MacGarvin, Andy Stirling, Jane Keys, Brian Wynne, and Sofia Guedes Vaz (eds). London: Earthscan, 2002.

Harrison, Gordon. *Mosquitoes. Malaria and Man: A History of the Hostilities Since 1880*. New York: E.P. Dutton, 1978.

Harrison, Mark. 'Disease, Diplomacy and International Commerce: The Origins of International Sanitary Regulation in the Nineteenth Century'. *Journal of Global History*, 1 (2006): 197–217.

——. '"Hot beds of disease"; Malaria and Civilization in Nineteenth-Century British India'. *Parassitologia*, 40 (1998): 11–18.

——. 'Public Health and Medical Research in India. c. 1860–1914'. diss. Oxford University, 1991.

——. 'Public Health and Medicine in British India: An Assessment of the British Contribution'. *Bulletin of the Liverpool Medical History Society*, 10 (1998): 32–48.

——. *Public Health in British India: Anglo-Indian Preventive Medicine, 1859–1914*. Cambridge: Cambridge University Press, 1994.

Haynes, Douglas. *Imperial Medicine: Patrick Manson and the Conquest of Tropical Disease*. Philadelphia: The University of Pennsylvania Press, 2001.

Hays, J. N. *Epidemics and Pandemics: Their Impacts on Human History*. Santa Barbara: ABC-CLIO, 2005.

Headrick, Daniel, R. *The Tools of Empire: Technology and Imperialism in the Nineteenth Century*. New York. Oxford University Press, 1981.

Henkin, Louis. *How Nations Behave: Law and Foreign Policy*. New York: Columbia University Press, 1979.

Hewa, Soma. *Colonialism. Tropical Disease and Imperial Medicine: Rockefeller Philanthropy in Sri Lanka*. Lanham: University Press of America, 1995.

Heymann, David, L. (ed.). *Control of Communicable Diseases Manual*. Washington: American Public Health Association, 2004.

Hooker, Claire. 'Sanitary Failure and Risk: Pasteurisation, Immunisation and the Logics of Prevention'. *Contagion: Epidemics. history and culture from smallpox to anthrax*. Alison Bashford and Claire Hooker (eds). Annandale: Pluto Press Australia, 2002 [2001].

Howard-Jones, Norman. *International Public Health between the Two World Wars: The Organizational Problems*. Geneva: World Health Organization, 1978.

——. *The Scientific Background of the International Sanitary Conferences, 1851–1938*. Geneva: World Health Organization, 1975.

Huber, Valeska. 'The Unification of the Globe by Disease? The International Sanitary Conferences on Cholera, 1851–1894'. *Historical Journal*, 49.2 (2006): 453–76.

Hyam, Ronald. *Britain's Imperial Century, 1815–1914: A Study of Empire and Expansion*. Lanham: Barnes and Nobles Books, 1993.

Inden, Ronald. *Imagining India*. London: Hurst & Company, 2000.

Iriye, Akira. *Cultural Internationalism and World Order*. Baltimore: The Johns Hopkins University Press, 1997.

Isaacs, Jeremy, D. 'D. D. Cunningham and the Aetiology of Cholera in British India, 1869–1897'. *Medical History*, 42 (1998): 279–305.

Jaggi, O. P. *Medicine in India: Modern Period*. New Delhi: Oxford University Press, 2000.

Jalal, Ayesha. 'Exploding Communalism: The Politics of Muslim Identity in South Asia'. *Nationalism. Democracy and Development: State and Politics in India*. Sugata Bose and Ayesha Jalal (eds). Delhi: Oxford University Press, 1997.

——. *Self and Sovereignty: Individual and Community in South Asian Islam since 1850*. London: Routledge, 2000.

Jamison, Dean, T. 'Cost-Effectiveness Analysis: Concepts and Applications'. *Oxford Textbook of Public Health*. Roger Detels, James McEwen, Robert Beaglehole and Heizo Tanaka (eds). Vol. 2. Oxford: Oxford University Press, 2002.

Jasanoff, Sheila. 'Beyond Calculation: A Democratic Response to Risk'. *Disaster and the Politics of Intervention*. Andrew Lakoff (ed.). New York: Columbia University Press, 2010.

Jeffery, Roger. 'Doctors and Congress: The Roles of Medical Men and Medical Politics in Indian Nationalism'. *The Indian National Congress and the Political Economy of India 1885–1985*. Mick Shepperdson and Colin Simmons (eds). Aldershot: Avebury, 1988.

——. *The Politics of Health in India*. Berkeley: University of California Press, 1988.

Jones, Margaret. *Health Policy in Britain's Model Colony: Ceylon, 1900–1948*. New Delhi: Orient Longman, 2004.

Kaminsky, Arnold, P. *The India Office, 1880–1910*. New York: Greenwood Press, 1986.

Keck, Frédéric. 'From Mad Cow Disease to Bird Flu: Transformations of Food Safety in France'. *Biosecurity Interventions: Global Health and Security in Question*. Andrew Lakoff and Stephen J. Collier (eds). New York: Columbia University Press, 2008.

Kent, Susan Kingsley. *Gender and Power in Britain, 1640–1990*. London: Routledge, 1999.

Klein, Ira. 'Death in India, 1871–1921'. *Journal of Asian Studies*, 32 (1973): 639–59.

——. 'Development and Death: Reinterpreting Malaria, Economics and Ecology in British India'. *Indian Economic and Social History Review*, 38.2 (2001): 147–79.

——. 'Malaria and Mortality in Bengal, 1840–1921'. *Indian Economic and Social History Review*, 9.2 (1972): 132–60.

——. 'Medicine and Culture in British India'. *Journal of Indian History*, 76–8 (1997–9): 122.

——. 'Plague. Policy and Popular Unrest in British India'. *Modern Asian Studies*, 22.4 (1988): 723–55.

——. 'Urban Development and Death: Bombay City, 1870–1914'. *Modern Asian Studies*, 20.4 (1986): 725–54.

Kohn, George C. (ed.). *Encyclopedia of Plague and Pestilence*. New York: Facts of File, 1995.

Kralev, Nicholas. 'KRALEV: US Warns Travelers of China's Flu Rules'. *Washington Times*, 29 June 2009. http://washingtontimes.com. Accessed 4 November 2009.

Kumar, Anil. 'The Indian Drug Industry under the Raj, 1860–1920'. *Health. Medicine and Empire: Perspectives on Colonial India.* Biswamoy Pati and Mark Harrison (eds). New Delhi: Orient Longman, 2001.

——, Anil. *Medicine and the Raj: British Medical Policy in India, 1835–1911.* Walnut Creek: Altamira Press, 1998.

Kumar, Deepak. 'Perceptions of Public Health: A Study in British India'. *Maladies. Preventives and Curatives: Debates in Public Health in India.* Amiya Kumar Bagchi and Krishna Soman (eds). Kolkata: Tulika Books, 2005.

Lakoff, Andrew. 'Introduction'. *Disaster and the Politics of Intervention.* Andrew Lakoff (ed.). New York: Columbia University Press, 2010.

Lakoff, Andrew and Stephen J. Collier (eds). *Biosecurity Interventions: Global Health and Security in Question.* New York: Columbia University Press, 2008.

Lal, Maneesha. '"The ignorance of women is the house of illness": Gender, Nationalism, and Health Reform in Colonial North Indian'. *Medicine and Colonial Identity.* Mary P. Sutphen and Bridie Andrews (eds). London: Routledge, 2003.

Leavitt, Judith Walzer. '"Typhoid Mary" Strikes Back Bacteriological Theory and Practice in Early Twentieth-Century Public Health'. *Isis,* 83.4 (1992): 608–29.

Lentzos, Filippa. 'The Pre-History of Biosecurity: Strategies of Managing Risks to Collective Health'. *Biosecurity: Origins, Transformations and Practices.* Brian Rappeprt and Chandré Gould (eds). Houndmills, Basingstoke: Palgrave Macmillan, 2009.

Litsios, Socrates. www.rockarch.org/publications/conferences/quinnipiac.php. Accessed 15 November 2008.

Löwy, Ilana. 'Producing a Trustworthy Knowledge: Early Field Trials of Anticholera Vaccines in India'. *Vaccinia, Vaccination, Vaccinology: Jenner, Pasteur and their Successors.* Stanley A. Plotkin and Bernardino Fantini (eds). Paris: Elsevier, 1996.

Ludden, David. 'Orientalist Empiricism: Transformations of Colonial Knowledge'. *Orientalism and the Postcolonial Predicament: Perspectives on South Asia.* Carol A. Breckenridge and Peter van der Veer (eds). Philadelphia: University of Pennsylvania Press, 1993.

Luhmann, Niklas. *Risk: A Sociological Theory.* New Brunswick: Aldine Transaction, 2002.

Lupton, Deborah. *The Imperative of Health: Public Health and the Regulated Body.* London: Sage Publications, 1995.

MacKenzie, John, M. 'Introduction'. *Imperialism and the Natural World.* John M. MacKenzie (ed.). Manchester: Manchester University Press, 1990.

MacLeod, Roy. 'Introduction'. *Disease, Medicine, and Empire: Perspectives on Western Medicine and the Experience of European Expansion.* Roy MacLeod and Milton Lewis (eds). London: Routledge, 1988.

Manning, P. K. 'Managing Risk: Managing Uncertainty in the British Nuclear Installations Inspectorate'. *Law and Policy,* 11.3 (1989): 350–69.

Martin, Andrew and Clifford Krauss. 'Pork Industry Fights Concerns Over Swine Flu'. *New York Times,* 28 April 2009. http://www.nytimes.com. Accessed 6 November 2009.

Mehta, Uday Singh. *Liberalism and Empire: A Study in Nineteenth-Century British Liberal Thought.* Chicago: The University of Chicago Press, 1999.

Metcalf, Thomas. *Forging the Raj: Essays on British India in the Heyday of Empire.* New Delhi: Oxford University Press, 2005.

——. *Ideologies of the Raj*. Cambridge: Cambridge University Press, 1994.
——. *Imperial Connections: India in the Indian Ocean Arena, 1860–1920*. Berkeley: University of California Press, 2007.
Miller, Talea. 'US Passes on Unlicensed H1N1 Vaccine Boosters, Despite Shortage'. *PBS NewsHour*, 9 November 2009. http://www.pbs.org. Accessed 20 December 2009.
Minault, Gail. *The Khilafat Movement: Religious Symbolism and Political Mobilization in India*. New York: Columbia University Press, 1982.
Minsky, Lauren. 'Pursuing Protection from Disease: The Making of Smallpox Prophylactic Practice in Colonial Punjab'. *Bulletin of the History of Medicine*, 83.1 (2009): 164–90.
Mishra, Saurabh. *Pilgrimage, Politics, and Pestilence: The Haj from the Indian Subcontinent, 1860–1920*. New Delhi: Oxford University Press, 2011.
Monath, Thomas, P. 'Yellow Fever'. *Vaccines*, 3rd edn. Stanley A. Plotkin and Walter A. Orenstein (eds). Philadelphia: W.B. Saunders Company, 1999.
Muraleedharan, V. R. '"Cinchona" Policy in British India: The Critical Early Years'. *Maladies. Preventives and Curatives: Debates in Public Health in India*. Amiya Kumar Bagchi and Krishna Soman (eds). Kolkata: Tulika Books, 2005.
——. 'Malady in Madras: The Colonial Government's Response to Malaria in the Early Twentieth Century'. *Science and Empire: Essays in India Context, 1700–1947*. Deepak Kumar (ed.). Delhi: Anamika Prakashan, 1991.
Mythen, Gabe. *Ulrich Beck: A Critical Introduction to the Risk Society*. London: Pluto Press, 2004.
Nathanson, Constance A. *Disease Prevention as Social Change: The State, Society, and Public Health in the United States, France, Great Britain, and Canada*. New York: Russell Sage Foundation, 2007.
National Intelligence Council. 'The Global Infectious Disease Threat and Its Implications for the United States'. January 2000.
——. 'The Next Wave of HIV/AIDS: Nigeria, Ethiopia, Russia, India, and China'. September 2002.
Neustadt, Richard, E. and Ernest, R. May. *Thinking in Time: The Uses of History for Decision Makers*. New York: The Free Press, 1986.
Neustadt, Richard, E. and Harvey, V. Fineberg. *The Epidemic That Never Was: Policy-Making and the Swine Flu Scare*. New York: Random House, 1982.
Nye, Edwin, R. and Mary, E. Gibson. *Ronald Ross: Malariologist and Polymath: A Biography*. London: Macmillan, 1997.
Ostergard, Jr., Robert, L. 'HIV/AIDS, State Capacity, and the Threat to National and International Security: A Theoretical Overview'. Robert L. Ostergard, Jr. (ed.) *HIV/AIDS and the Threat to National and International Security*. Houndmills, Basingstoke: Palgrave Macmillan, 2007.
Ozcan, Azmi. *Pan-Islamism: Indian Muslims. The Ottomans and Britain, 1877–1924*. Leiden: Brill, 1997.
Packard, Randall, M. *The Making of a Tropical Disease: A Short History of Malaria*. Baltimore: The Johns Hopkins University Press, 2007.
Palit, Chittabrata. *Scientific Bengal: Science, Technology, Medicine and Environment under the Raj*. Delhi: Kalpaz Publications, 2006.
Pande, Ishita. *Medicine, Race and Liberalism in British Bengal: Symptoms of Empire*. London: Routledge, 2010.

Patel, Sapna. 'Responses to an Epidemic: The Attitudes of the Colonial State and the Indian Press to the Plague in Bombay: September 1896 to December 1897'. diss. Wellcome Institute, 2005.

Pati, Biswamoy. and Mark Harrison. 'Introduction'. *Health. Medicine and Empire: Perspectives on Colonial India*. Biswamoy Pati and Mark Harrison (eds). New Delhi: Orient Longman, 2001.

Petersen, Alan and Deborah Lupton. *The New Public Health: Health and Self in the Age of Risk*. St. Leonards, Australia: Allen and Unwin, 1996.

Peterson, Susan. 'Human Security, National Security, and Epidemic Disease'. Robert L. Ostergard, Jr. (ed.) *HIV/AIDS and the Threat to National and International Security*. Houndmills, Basingstoke: Palgrave Macmillan, 2007.

Pidgeon, Nick. 'Risk, Uncertainty and Social Controversy: From Risk Perception and Communication to Public Engagement'. *Uncertainty and Risk: Multidisciplinary Perspectives*. Gabriele Bammer and Michael Smithson (eds). London: Earthscan, 2008.

Plant, Aileen, J. 'When Action Can't Wait: Investigating Infectious Disease Outbreaks'. *Uncertainty and Risk: Multidisciplinary Perspectives*. Gabriele Bammer and Michael Smithson (eds). London: Earthscan, 2008.

Porter, Bernard. *The Lion's Share: A Short History of British Imperialism, 1850–2004*. Harlow: Pearson, 2004.

Porter, Dorothy. *Health. Civilization and the State: A History of Public Health from Ancient to Modern Times*. London: Routledge, 1999.

Power, Helen, J. *Tropical Medicine in the Twentieth Century: A History of the Liverpool School of Tropical Medicine, 1898–1990*. London: Kegan Paul, 1999.

Prakash, Gyan. *Another Reason: Science and the Imagination of Modern India*. Princeton: Princeton University Press, 1999.

Price-Smith, Andrew, T. *Contagion and Chaos: Disease, Ecology, and National Security in the Era of Globalization*. Cambridge: MIT Press, 2009.

——. 'Ghosts of Kigali: Infectious Disease and Global Stability at the Turn of the Century'. Andrew T. Price-Smith (ed.). *Plagues and Politics: Infectious Disease and International Policy*. Houndmills, Basingstoke: Palgrave Macmillan, 2001.

Pugh, Martin. *The Making of Modern British Politics, 1867–1945*. Oxford: Blackwell Publishers, 2002 [1982].

Ray, Kabita. 'Press and the Problem of Medical Relief in Colonial Bengal, 1921–1947'. *Maladies, Preventives and Curatives: Debates in Public Health in India*. Amiya Kumar Bagchi and Krishna Soman (eds). New Delhi: Tulika Books, 2005.

Raychaudhuri, Tapan. 'India, 1858 to the 1930s'. *The Oxford History of the British Empire: Volume V: Historiography*. Robin W. Winks (ed.). Oxford: Oxford University Press, 1999.

Rodriguez, William, R. 'Interview with Robert C. Bollinger and William R. Rodriguez'. http://www.globalur.com/node/64. Accessed 8 December 2008.

Rogers, Leonard. 'Progress in Control of Cholera in India by Inoculation of Pilgrims'. *British Medical Journal*, 2.4796 (1952): 1219–20.

Rosen, George. *A History of Public Health*. New York: MD Publications, 1958.

Rosenberg, Charles, E. 'Anticipated Consequences: Historians, History, and Health Policy'. *History and Health Policy in the United States: Putting the Past Back In*. Rosemary A. Stevens, Charles E. Rosenberg, and Lawton R. Burns (eds). New Brunswick: Rutgers University Press, 2006.

——. 'Banishing Risk: Continuity and Change in the Moral Management of Disease'. *Morality and Health*. Allan M. Brandt and Paul Rozin (eds). New York and London: Routledge, 1997.

——. 'Cholera in 19th-Century Europe: A Tool for Social and Economic Analysis' in *Explaining Epidemics*. Cambridge: Cambridge University Press, 1992.

——. 'Framing Disease' in *Explaining Epidemics and Other Studies in the History of Medicine*. Cambridge: Cambridge University Press, 1992.

——. 'Pathologies of Progress: The Idea of Civilization as Risk'. *Bulletin of the History of Medicine*, 72.4 (1998): 714–30.

——. *The Cholera Years: The United States in 1832, 1849, and 1866*. Chicago: The University of Chicago Press, 1987 [1962].

Rothermund, Dietmar. *India in the Great Depression, 1929–1939*. New Delhi: Manohar, 1992.

Sachs, Jeffery. 'Macroeconomics and Health: Investing in Health for Economic Development'. Geneva: World Health Organization, 2001.

——. http://www.wilsoncenter.org. 16 April 2002. Accessed 8 December 2008.

Said, Edward W. *Orientalism*. New York: Vintage Books, 1978.

Samanta, Arabinda. *Malarial Fever in Colonial Bengal, 1820–1939: Social History of an Epidemic*. Kolkata: Firma KLM, 2002.

Samson, Jane. *Race and Empire*. Harlow: Pearson Education LimitEd, 2005, p. 85.

Sarkar, Sumit. *Modern India: 1885–1947*. London: Macmillan Press, 1989.

Sen, Sudipta. *Distant Sovereignty: National Imperialism and the Origins of British India*. New York: Routledge, 2002.

Sibbons, Maureen. 'Cholera and Famine in British India, 1870–1930'. *Papers in International Development*. No. 14. Swansea: Centre for Development Studies, 1995.

Sinha, Sandeep. *Public Health Policy and the Indian Public: Bengal 1850–1920*. Calcutta: Vision Publications, 1998.

Snowden, Frank M. *The Conquest of Malaria: Italy, 1900–1962*. New Haven: Yale University Press, 2006.

Stepan, Nancy. 'The Interplay between Socio-Economic Factors and Medical Science: Yellow Fever Research. Cuba and the United States'. *Social Studies of Science*, 8.4 (1978): 397–423.

Stern, Alexandra Minna. 'Yellow Fever Crusade: US Colonialism, Tropical Medicine, and the International Politics of Mosquito Control, 1900–1920'. *Medicine at the Border: Disease, Globalization and Security, 1850 to the Present*. Alison Bashford (ed.). New York: Palgrave Macmillan, 2006.

Stern, Alexandra Minna, and Howard Markel. 'International Efforts to Control Infectious Diseases, 1851 to the Present'. *JAMA*, 292 (2004): 1474–9.

Stern, Paul, C. and Harvey, V. Fineberg (eds). *Understanding Risk: Informing Decisions in a Democratic Society*. Washington, DC: National Academy Press, 1996.

Stevens, Rosemary, A. 'Introduction'. *History and Health Policy in the United States: Putting the Past Back In*. Rosemary A. Stevens, Charles E. Rosenberg, and Lawton R. Burns (eds). New Brunswick: Rutgers University Press, 2006.

Stoler, Ann Laura. and Frederick Cooper. 'Between Metropole and Colony: Rethinking a Research Agenda'. *Tensions of Empire: Colonial Cultures in a Bourgeois World*. Frederick Cooper and Ann Laura Stoler (eds). Berkeley: University of California Press, 1997.

Summers, William, C. 'Cholera and Plague in India: The Bacteriophage Inquiry of 1927–1936'. *Journal of the History of Medicine and Allied Sciences*, 48.3 (1993): 275–301.

'Tensions escalate in Hong Kong's swine-flu hotel'. *Taipei Times*, 3 May 2009. http://www.taipeitimes.com. Accessed 5 November 2009.

Thompson, Donald, F. and Renata, P. Louie. 'Cooperative Crisis Management and Avian Influenza: A Risk Assessment Guide for International Contagious Disease Prevention and Risk Mitigation'. March 2006. http://www.ndu.edu/ctnsp/Def_Tech. 28–29. (Accessed 3 April 2009.)

Tierney, Kathleen J. 'Toward a Critical Sociology of Risk'. *Sociological Forum*. 14.2 (1999): 215–42.

Todd, Brian. and Taylor Gandossy. 'China Quarantines US School Group Over Flu Concerns'. *CNN*, 28 May 2009. http://www.cnn.com. Accessed 4 November 2009.

Tomlinson, B. R. *Political Economy of the Raj, 1914–1947: The Economics of Decolonization in India*. London: Macmillan Press, 1979.

Towers, Sherry. and Zhilan Feng. 'Novel H1N1: Predicting the Course of a Pandemic'. http://www.stat.purdue.edu. Accessed 20 November 2009.

Tuck, Richard. *The Rights of War and Peace: Political Thought and the International Order from Grotius to Kant*. Oxford: Oxford University Press, 1999.

Tucker, Jonathan, B. 'Updating the International Health Regulations'. *Biosecurity and Bioterrorism: Biodefense Strategy, Practice, and Science*, 3.4 (2005): 338–47.

Van Vleck, Jenifer. 'The Logic of the Air: Visions of Aviation and Empire, 1938–1945'. Harvard University. Cambridge, 16 March 2007.

Verma, D. N. *India and the League of Nations*. Patna: Bharati Bhawan, 1968.

Wailoo, Keith. *Drawing Blood: Technology and Disease Identity in Twentieth-Century America*. Baltimore: Johns Hopkins University Press, 1997.

Washbrook, David. 'Orients and Occidents: Colonial Discourse Theory and the Historiography of the British Empire'. *The Oxford History of the British Empire: Volume V: Historiography*. Robin W. Winks (ed.). Oxford: Oxford University Press, 1999.

——. 'The Rhetoric of Democracy and Development in Late Colonial India'. *Nationalism. Democracy, and Development*. Sugata Bose and Ayesha Jalal (eds). Delhi: Oxford University Press, 1997.

Watts, Sheldon. 'British Development Policies and Malaria in India 1897–c. 1929'. *Past and Present*, 165 (1999): 141–81.

Weir, Lorna. and Eric Mykhalovskiy. *Global Public Health Vigilance: Creating a World on Alert*. New York: Routledge, 2010.

Whitcombe, Elizabeth. 'The Environmental Costs of Irrigation in British India: Waterlogging, Salinity, Malaria'. *Nature. Culture. Imperialism: Essays on the Environmental History of South Asia*. David Arnold and Ramachandra Guha (eds). Delhi: Oxford University Press, 1995.

Wilson, Kathleen. *A New Imperial History: Culture. Identity and Modernity in Britain and the Empire, 1660–1840*. Cambridge: Cambridge University Press, 2004.

Woolls, Daniel. 'Swine Flu Cases in Europe; Worldwide Travel Shaken'. *Taiwan News*, 28 April 2009. http://www.etaiwannews.com. Accessed 5 November 2009.

Worboys, Michael. 'British Colonial Medicine and Tropical Imperialism: A Comparative Perspective'. *Dutch Medicine in the Malay Archipelago, 1816–1942*. A.M. Luyendijk-Elshout (ed.). Amsterdam: Rodopi B.V., 1989.

——. 'Germs, Malaria and the Invention of Mansonian Tropical Medicine: From "Diseases in the Tropics" to "Tropical Diseases"'. *Warm Climates and Western Medicine: The Emergence of Tropical Medicine, 1500–1900*. David Arnold (ed.). Amsterdam: Rodopi B.V., 1996.

——. 'Manson. Ross. and Colonial Medical Policy: Tropical Medicine in London and Liverpool, 1899–1914'. *Disease, Medicine, and Empire: Perspectives on Western Medicine and the Experience of European Expansion*. Roy MacLeod and Milton Lewis (eds). London: Routledge, 1988.

——. 'The Discovery of Colonial Malnutrition between the Wars'. *Imperial Medicine and Indigenous Societies*. David Arnold (ed.). Manchester: Manchester University Press, 1988.

——. 'The Imperial Institute: The State and the Development of the Natural Resources of the Colonial Empire, 1887–1923' and John M. MacKenzie. 'Experts and Amateurs: Tsetse, nagana and Sleeping Sickness in East and Central Africa'. *Imperialism and the Natural World*. John M. MacKenzie (ed.). Manchester: Manchester University Press, 1990.

——. *Spreading Germs: Disease Theories and Medical Practice, 1865–1900*. Cambridge: Cambridge University Press, 2000.

Wylie, Diana. 'Disease, Diet, and Gender: Late 20th Century Perspectives on Empire'. *The Oxford History of the British Empire: Volume V: Historiography*. Robin W. Winks (ed.). Oxford: Oxford University Press, 1999.

Zacher, Mark, W. and Tania J. Keefe. *The Politics of Global Health Governance: United by Contagion*. Houndmills, Basingstoke: Palgrave Macmillan, 2008.

Zylberman, Patrick. 'Civilizing the State: Borders, Weak States and International Health in Modern Europe'. *Medicine at the Border: Disease, Globalization and Security, 1850 to the Present*. Alison Bashford (ed.). New York: Palgrave Macmillan, 2006.

# Index

AIDS, *see* HIV/AIDS
All-India Sanitary Conferences, 66, 72
Ayurvedic medicine, 22

Beck, Ulrich, 1, 7
Bengal, *see* Governments
Bihar, *see* Governments
Biosecurity, 9, 136
 public health preparedness, 2, 9,
  127, 136
Bombay Haj Committee, 61
Bombay, *see* Governments
British Empire
 Africa, 133
 Burma, 46–7
 economy, 51–3
 trade, 10
 labor, 52–3
 structure of, 19–20, 140
 Whitehall, 39, 56
 *see also* Colonialism

Caliph, 57
Cholera, 12, 26–31, 38, 50, 75
 epidemic, 29–30
 pandemic, 26
 vaccination, 59–62, 75
Colonialism
 ideologies, 4–5, 14, 15–17, 76, 155
 development policies, 13
 politics of collaboration, 49, 55,
  57–8, 217
 rhetoric, 14, 26, 63, 69, 80, 97, 113
 theories of, 4–5, 16–17, 20, 155
Colonial Office, 116, 132, 150
Curzon, Lord, 41

Devolution, 74, 80, 98–9, 104, 127,
 154
Disease
 endemic, 50–1, 68–70, 74
 etiologies, 58, 65
 *see under* individual names

Dispensaries, 17
Dyarchy, 98

Egypt, 134
 Khedive, 30–1
Elgin, Lord, 32, 33, 35, 57
Embargoes, 31–2, 33, 51
Epidemic Diseases Act, *see* Public
 Health, legal acts
Epidemiology, 7, 26, 50, 61, 65, 82–3,
 121, 135
Europe, 8, 15, 18, 26–7, 26–49

Far Eastern Association of Tropical
 Medicine, 19, 67

Giddens, Anthony, 7
Great Depression, 107
Government of India Act 1919, 66,
 73, 74, 101–2, 116, 154
Government of India Act 1935, 137
Governments
 of Bengal, 21, 64–5, 86, 89, 101, 127
 of Bihar and Orissa, 111–12
 of Bombay, 21, 32, 126
 of Ceylon, 45–6
 of Madras, 21, 44–5, 47, 101
 of Northwest Provinces, Oudh, 55
 of Punjab, 62
 of Seychelles, 53

Haffkine, Waldemar, 64, 69–70
*Hajj, see* Pilgrimage
Hamilton, Lord, 32, 33
Hedjaz, *see* Pilgrimage
Historiography, 4–5, 8, 23
HIV/AIDS, 1–2
 PEPFAR, 2
Hospitals, 17
Hygienists, 8

India Office, 20, 147
Indian Medical Service, 66, 70, 91